Copyright Date 2016-2021
Hardcover ISBN 978-1-932113-63-1
Paperback ISBN 978-1-932113-62-4
Published by Lauric Press

www.ForceNecessary.com

All rights reserved. Reproduction or electronic file sharing or distribution without written permission from the author is specifically prohibited.

The author, publishers or sellers will not and do not assume any responsibility for the use or misuse of information contained in this book. The one and only purpose of this publication is to provide information for the purpose of self-defense survival against illegal aggression. Use the information only when it is morally, legally and ethically appropriate to do so.

Other Martial Training Book Titles by W. Hock Hochheim
Fightin' Words-The Psychology and Physicality of Fighting,
Knife/Counter-Knife Combatives
Impact Weapon Combatives

TABLE OF CONTENTS

Introduction	Page 5
Mission Overview	Page 7
Segment 1: Universal Basic Concepts	Page 9
Segment 2: The Stand-Up Footwork	Page 37
Segment 3: The Knee Maneuvers	Page 61
Segment 4: Standing to Knee to Ground Maneuvers	Page 73
Segment 5: Crawling Maneuvers	Page 83
Segment 6: Universal Grounded Maneuvers	Page 91
Segment 7: Universal Get-Up Maneuvers	Page 107
Segment 8: Obstacle Courses	Page 113
Segment 9: Your Personal, Class, or Seminar Workout List	Page 121

Introduction - "Getting Into Position To..."

"I have been an adult-lifetime martial artist, though "artist" is not the best term for my obsession, but "artist" is, however, an easy word for regular people to recognize. I have worked my way to black belt in several systems. I have toiled in the footwork of martial systems and boxing, kick boxing, stick, knife and gun fighting.

I have also been in the US Army Military Police and have retired from decades of police work in the US, most of that time serving as a detective. I followed that up with years in private investigation. At the time of this writing and for the last 23 years, I do and have taught citizens, police and military units in 12 different countries, recording their needs, their successes and failures.

I have come to understand that "footwork," is not just about moving around in a boxing ring or MMA octagon. It's not just about rolling around on a soft mat, but rather about traversing all kinds of real-world surfaces inside and outside of homes and buildings in urban, suburban and rural areas, in all kinds of weather, day and night, while in real clothes and carrying all sorts of gear. Unarmed or armed with sticks, knives, pistols and long guns. It's about fighting while standing, seated, kneeling and on the "ground."

Fighting footwork and maneuvers. It is a combination of sports and obstacle courses. And by obstacle courses I mean ones designed to replicate real environments, not science fiction-esque, torture runs to see how muddy you can get.

If you can walk and if you can run, you have a good foundation for footwork. If you can roll around on the ground, do some sit-ups, some push-ups and squats, you can maneuver on the ground.

As in life's maneuvers, there are borders, and the training mission of this publication has a border. This book just focuses on moving around, moving your body into positions before, during and after some sort of fight. Avoidance positions. Defense positions. Attack positions. The actual moves of fighting like striking, kicking, blocking, stick work, knife work and shooting are other topics covered in other studies.

This book, this manual, is the culmination of more than 40 years of observation, education and experience, a holistic collective of hand, stick, knife footwork and ground maneuvers, with workout lists and outlines in the last segment anyone can use to improve their performances, or use in class or in a seminar.

Mission Overview

"The infantry learns to love the ground." This is an old military expression. After being on the receiving end of gunfire and lobbed explosives, a ducking troop instinctively learns to see where he or she can find the best cover. If there is none, one penetrates the ground with entrenching tools, even with desperate fingers to dig down deep if that is all one has.

Soon after, when this veteran troop walks across the next potential hot zone, they reflexively study the very lay of the land ahead. One now spots even the slightest, natural incline, decline or man-made structure that might save their life. Cover or at least concealment.

The Infantry must learn to love the land, and close quarter fighters with or without weapons, must learn to see and feel the floors and ground they will do battle on and have the savvy, agility and strength to overcome the variables of weather, surface and space.

Many innocently think that "fighting footwork" comes from the boxing ring, relying on movements like the shuffle-step, the rocker and so forth. All martial arts have some pattern laid out on the mat upon which to dart back and forth, side to side. I prefer the clock pattern. But, no matter the pattern, consummate trainees learn to cover air, water and land.

Exclusive of parachuting and scuba, for any citizen, enforcement officer or soldier, covering turf/land is done three ways, by crawling, walking and running over:
    1: urban
    2: suburban
    3: rural terrain (hills, plains, desert, jungle, mountains, etc.)

These terrains are defined as the outsides and insides of the vast variety of man-made buildings and structures, and unpopulated, under-populated, populated and over-populated areas. All this turf is traversed in differing kinds of weather and lighting. When in armed conflicts, under three shooting predicaments:
    1: no fire conditions
    2: light fire conditions
    3: heavy fire conditions

Once past close quarters ring footwork, warriors traverse terrain, often fast. Nothing replaces running regularly to facilitate this goal. It builds wind, endurance and spirit. Experts will say that a regular regimen of jogging and wind sprints is the best combination. Treadmills are nice, but I believe you must run outdoors, and in all kinds of weather, to maximize your potential. Eventually you must exercise in the very environments of your next mission. Reduce the abstract.

Combative movement is an athletic endeavor. Footwork is a lot like walking and running, but it can be advanced to a more tactical walking and running, especially when carrying weapons. Your survival may hinge upon your ability to perform big and small footwork. The foundation for combat footwork comes from four main sources:

1: walking and running footwork.
2: sports footwork.
3: obstacle courses footwork.
4: applications of ground fighting maneuvers.

You need:
1: proper foot wear, socks and foot care.
2: jogging.
3: wind sprints.
4: arm, torso and leg strength training.
5: footwork floor patterns.
6: sport-related footwork practice.
7: obstacle courses.
8: ground work.
9: all empty handed and "while holding:"
   - edged weapons.
   - impact weapons.
   - firearms.

Being in better shape will also help you control your heart rate and related adrenaline.

Segment 1: Universal Basic Concepts

There are basic, universal principles of movement that the citizen, the enforcer, the soldier, the hand, stick, knife, and gun fighter must know and apply these to successfully approach or escape, defeat and, or, survive an opponent. The essential principles mentioned here are a must-know, important collection.

Basics 1: The Ready Stance
Martial systems obsess about fighting stances, often molding statues of frozen positions, all of which move the instant a fight starts. Fighting is about balance and power in motion, not worshipping a still photograph of a fighting position.
- (1) Knees somewhat bent.
- (2) Body bladed as needed, or hips straight and feet staggered.
- (3) Hands up somewhere in the "window of combat" that rectangle loosely defined from face to belt line.
- (4) From here, personal fighting "styles" are developed to suit the person and the customized training.

Basics 2: Physical Balance.
Balance refers to the ability to maintain equilibrium and to remain in a stable, upright position, or a controlling position on the ground. A fighter must maintain his or her balance as much as possible to defend oneself and to launch an effective attack. The fighter must understand two aspects of balance in a struggle:
- (1) How to move the body to keep or regain balance. A fighter develops balance through experience, but usually keeps the feet about shoulder-width apart and knees flexed. Lower ones center of gravity to increase stability.
- (2) How to exploit weaknesses in his opponent's balance. Experience also gives the hand-to-hand fighter a sense of how to move his body in a fight to maintain ones balance while exposing the enemy's weak points.

Basics 3: Mental Balance
The successful fighter must also maintain a mental balance. He or she must not allow fear or anger to overcome ones ability to concentrate or to react instinctively. I suggest fear management, anger management and pain management programs. These subjects are discussed in other books, not this one which is about footwork.

Basics 4: Position
The need for exact positioning is situational in the hand, stick, knife, gun worlds. Position refers to the location of the fighter-defender in relation to his or her opponent. A vital principle when being attacked is for the defender to move ones body to a safe, solid or springboard position. Movement to an advantageous position requires accurate timing, distance perception, and the savvy to know where that best position is. Sometimes it's instinctual. Sometimes trained. Sometimes a fortunate accident.

Basics 5: Moving with Time and Timing

A fighter must be able to perceive the best time to move to an advantageous position in an attack. If one moves too soon, the enemy will anticipate ones movement and adjust the attack. If the fighter moves too late, the enemy will strike. Similarly, the fighter must launch an attack or counterattack at the critical instant when the opponent is the most vulnerable.

When initiating a fake, the fighter must wait for a reaction. You must get a reaction to the fake, before making the real move. His reaction will be slower for untrained or dulled/diminished fighters.

Basics 6: Distance

Distance is the relative distance between the positions of opponents. A fighter should position his or herself where distance is to their advantage, as customized in the hand, stick, knife, gun worlds. Danger zones may begin from within sniper range on down to nose-to-nose.

Modern police trainers advise that if a person suddenly becomes suspicious, the officer should step back to, and remain "two giant steps and a lunge" away from the person. People should still be aware that a motivated attacker can still close that gap "in a split second," but perhaps one might better detect the attack.

Basics 7: Momentum

Defined as - the quantity of motion of a moving body, measured as a product of its mass and velocity. Think about football and rugby players crashing into each other. Consider their size, the speed, all mixed with momentum. Momentum is the tendency of a body in motion to continue in the direction of motion unless acted on by another force. Body mass in motion develops momentum. The greater the body mass or speed of movement, the greater the momentum. Therefore, a fighter must understand the effects of this principle and apply it to his advantage. Military manuals suggest:

(1) The fighter can use the opponent's momentum to an advantage-that is, he or she can place the opponent in a vulnerable position by using their momentum against him.

    (a) The opponent's balance can be taken away by using his own momentum.

    (b) The opponent can be forced to extend further than he expected, causing him to stop and change his direction of motion to continue his attack.

    (c) An opponent's momentum can be used to add power to a fighter's own attack or counterattack by combining body masses in motion.

(2) The fighter must be aware that the enemy can also take advantage of the principle of momentum. Therefore, the fighter must avoid placing his or herself in an awkward or vulnerable position, and avoid extending too far.

Basics 8: Awkward Positions
Another old military adage is that "combat is fighting from awkward positions." Think about the unusual and strange cover and concealment positions that occur indoors and outdoors. Think about moving from open spot A to open spot B or cover spot B. Think about hand to hand combat inside a house, inside a room full of furniture, or fighting while sliding down a muddy hill in a rainstorm. Crisis rehearse and prepare your mind and body for the common and the awkward.

Basics 9: Leverage
A fighter moves in and around to get and use leverage in close quarters by using the natural movement of his or her body to place ones opponent in a position of unnatural movement. Leverage often comes from footwork and ground positioning.

Basics 10: Make the Sports Connection
Much fighting footwork and maneuvering already resembles the footwork of many common sports like football, basketball, soccer, etc. Make sure you remind yourself and your practitioners to recall sports footwork methods they already know. This will speed up the learning process.

Basics 11: Clear All Snags
Limit your gear and clothing from snags and being hooked. Study yourself and your team's attires to see what on you can knock things over, snatch, catch, snag, make noise, cause refections or otherwise reveal your position, slow you down or trip you up. (Also, ever have to ground fight with a wallet in your pocket? How about with a badge in that wallet?).

Basics 12: The Changing, Challenging Surfaces
One of my more "memorable" ground fights was sliding down a steep, grass and mud hill, in a hard rain outside a hospital fighting a crazy guy bound for a mental ward. Another worst case one was fighting with dunk, agitated soldiers, a platoon vs. a platoon on gravel picnic grounds. We hit the ground, and I felt every stone. We fight on and, or in:

    (1) Various indoor surfaces like tile, carpet, linoleum, indoor cement floors. (Have I left some surface out? Probably.)

    (2) Various outdoor surfaces like astro-turf, dirt, grass, mud, gravel, muck, asphalt, sidewalks.

    (3) Weather. Rain. Sleet. Snow. Ice. Blistering egg-frying hot. Super cold.

    (4) Angles. Flat to very steep.

    (5) Stuff around you. Cars. People. Furniture. Trees. Bushes...am I leaving something out? Oh yes. Think of all the stuff around you at any given time and place. You could be fighting there. Standing or grounded.

Basics 13: Smooth Walking
There are times when a citizen/enforcer/soldier/Marine must walk keeping their pistol or long gun sights on an enemy or a potential enemy location. One may have to shoot while advancing. Yet, walking will interfere with this process. As the body moves the sight picture sways and jumps. This is always a challenge. Running will be worse. Some age-old, veteran advice and training tips are...

(1) Walking with "wide" steps will cause a side-to-side sway of the gun. Plus, wide knees technically offer a bigger target.

(2) Walking with a narrow knee stride, will cause an up-and-down motion of the sight picture. This problem has been more desirable than the side-to-side sway by professionals.

(3) Professionals advise the "Groucho Marx" walk, a lower bent knees position, knees closer in stride, with elbows down.

(4) Going forward? Roll heel to toe. Going backward? Toe to heel.

(5) The need for speed may interfere with these precision steps.

Note: These subjects often reveal the need for quiet when moving, as well. It can be very hard to run, walk or crawl and maintain a low level of noise.

Basics 14: Ghost Walking. Ghost Crawling
"Ghost walking" has come to mean several different things in modern slang and culture. Ghosts was even a nickname for Delta Force. But ghost walking (or crawling) is an old term for stealth and also means moving, running, jogging, walking and crawling as quietly as possible. Approaching someone or someplace as quietly as possible leads to a good ambush and element of surprise. You are worried about noise discipline and visual discipline. You are concerned with:

(a) your mission and the details of its location.

(b) the sounds you make while moving.

(c) the sounds (like rattling?) your shifting uniform gear makes.

(d) identify your items that might cause glare and, or refect light.

(e) selecting the surreptitious path and not the obvious path of least resentience.

(f) the back ground of your silhouette as seen by the enemy.

(g) making foot or body contact with bushes, trees, plant life.

(h) making contact with manmade objects, like fences, rusty gates.

(i) tripping and, or falling.

(j) move within local sound bursts, like cars, planes, engines, etc.

(k) avoid grunts and words.

(l) set your foot down carefully and deliberately when walking.

(m) set your hand down carefully and deliberately when crawling.

(n) dragging your body will create noise, why many use the "high crawl."

(o) some military experts suggest heel first plant then toe, some even suggest toe then heel.

(p) break up the sound silhouette by not moving with a detectable rhythm.

(q) are there any food, tobacco or gear smells?

(r) night vision devices that operate by heat detection, picking up on you or where you recently "smeared," such as a rock or man made object.

(s) an old burglar trick is to remove shoes and socks, put on the shoes and put the socks over them to cushion the sounds of shoes on certain "noisy" floors.

Universals - A Word About Momentum and Explosiveness

"The quantity of motion of a moving body, measured as a product of its mass and velocity." That's what the good books of physics say. I do think we all have an idea about what momentum means. Mass and velocity. Just watch an American football game or video of a car crash.

Explosive footwork is a term usually attached to sports. Getting where you need to go as fast an efficiently with the power needed to accomplish the goal. Just about every competitive sport worries about agility, speed and needed force/power.

"HIKE!"

"BANG!" (the starters gun)

There's two big examples of explosive power and time spent training to maximize the instant. It would be hard to translate the training drills viewable in clips, on the internet into the written word. Look them up on the web and do and take what you think you need from them.

Think of that power. Now, think about passing aside that power in a fight. Lots of martial systems do think and worry about it. They train people to pass that sort of fully committed, or over-committed rush, off to the sides with opposing footwork assisted by circling or pushing hands and arms. They even do it successfully in football. In any given American football game, there is a whole lot of Aikido and Filipino Hubad going on, with full speed and force.

In the 70s, 80s and 90s, many noted Jeet Kune Do people made a lot of money teaching "how-to-pass" martial drills to professional football players and teams, notably the infamous "Hubad" drill. Get them by you! Get to the outside!

The moral of the story is - this explosive, rushing energy footwork, this excessive momentum footwork can be used against you. Not in a sport, footrace, but in contact sports and certainly in a fight. Perhaps this is where wisdom, agility and experience come in. But, even the best of the best can be shoved aside at times, as evidenced consistently in the real world of fighting and sports.

Here JKD Legend Tim Tackett confounds All Star Randy White with a "Hubad' martial hand drill.

Universals - The Combat Clock for Movement

The clock points. I was familiar with military terminology and concepts from my Army experience. If you were on a foot patrol and the point man suddenly shouted, "Enemy at 2 o'clock!" Everyone would instantly look in that direction. The same for pilots - who also have both a vertical clock and a horizontal clock. "12 o'clock high!" Simple. Quick. Effective. Unforgettable.

Yet, scores of differing police and martial arts training systems are not clock-based. Their elaborate weapon, angles of attack were disjointed and forgettable, a major problem in this frustrating rat race of systemologies, and the various lines of attack protocols each used for hand, stick, knife and gun tactics. Each try to out-smart or out-do the other, rather than focus in on maximizing education. The worst, in my opinion, are the two extremes - the over-simplified and the over-complicated. I began to ask myself, how are all these directions/angles of combat the same? It became clear that attacks universally come in from the center, high or low, or right or left sides, whether standing or on the ground.

No matter the weapon, the angles/directions are the same. I returned with trust to the simple, military "combat" clock. The clock face is an imprinted image in our minds since early childhood. The simple angle of attack pattern is right on your wrist, work or play. I discovered, or better said, I re-discovered the simple, military clock method as a training foundation. Stand it up or lay it down, you have an unforgettable pattern to teach, memorize and work from.

Basic Combat Clock Training:
12 o'clock from axis to above/top.
3 o'clock from axis to the right side.
6 o'clock from axis to below/bottom.
9 o'clock from axis to the left side.
Axis point is the center.

Advanced Combat Clock Training:
From the axis center point out to all 12 numbers of the clock. This 12 option offers more precision training if needed.

The Combat Clock is used in the following ways, to name just a few:
Learn hand, stick, knife and gun manipulation and solo command and mastery skills.
    Maneuvering - organize attack and defense footwork if laid horizontal.
    Target spotting - direct fire/locate enemies with a vertical and horizontal clock.
    Delivery system - use to deliver angles of attack.
    Blocking system.
    Organize attack striking, hooking/slashing strikes if set vertically.
    Organize attack shooting/stabbing/thrusting points if set vertically.
    Organize defensive moves if set vertically.
    Coordinate mission timing.

I have taught thousands and thousands of people from utter novices to experts, from cadets and rookie cops to vets and martial arts black belts, from all over the world, in these last 35 years, and I can get them to interact with each other in mere moments by using this simple basic Combat Clock number/point format.

In the end of this book you will find a collection, a list of unarmed and armed exercises you can use, utilizing these movements with the Combat Clock.

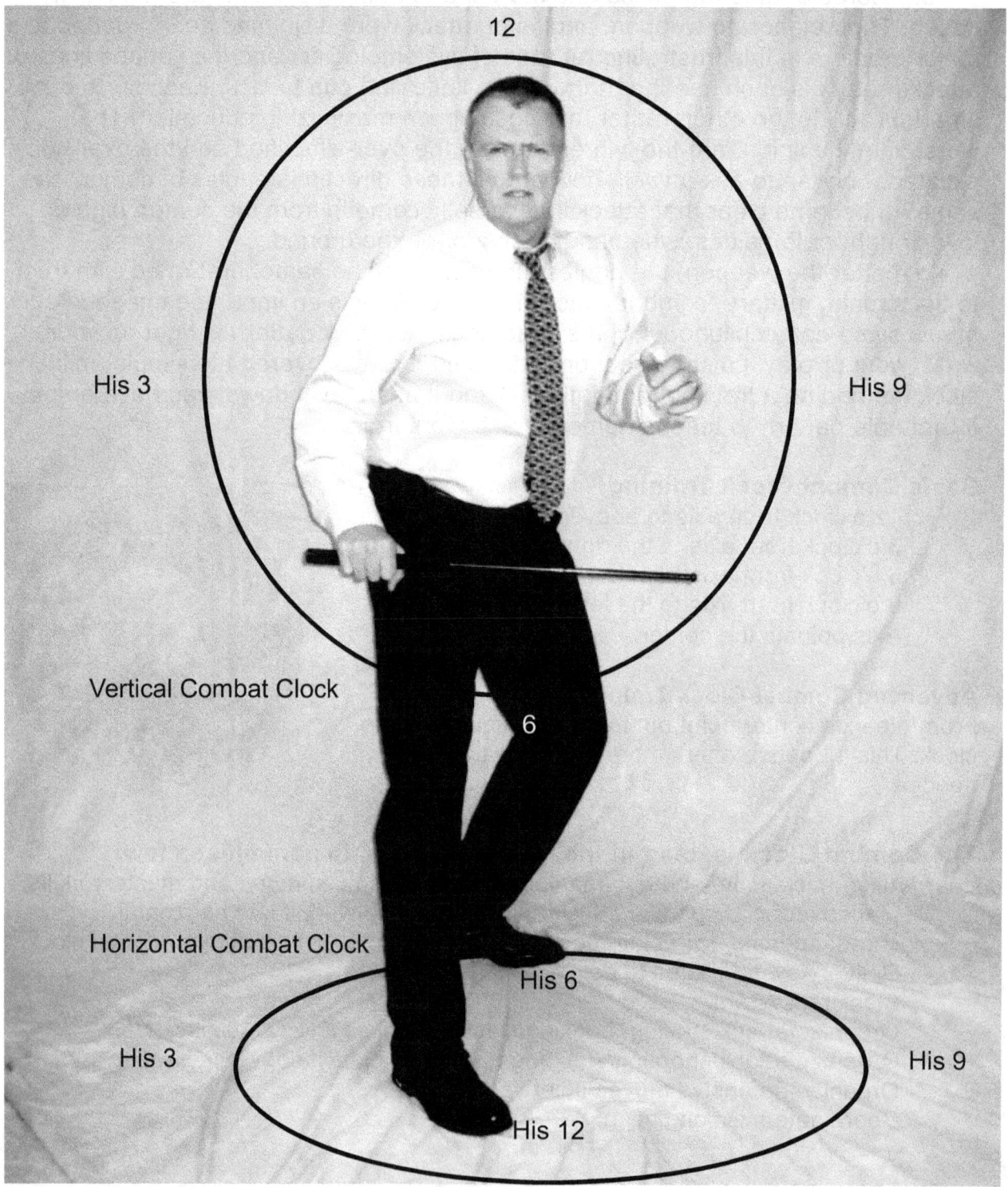

Universals - Where Footwork and Maneuvering Exists

Excluding missiles and hand grenades, in the hand, stick, knife, gun world, the skills in footwork and maneuvering run from sniper range (some 1,000 yards?) on down to face-to-face on the ground, and all that turf in between. Use this list to plan training and options.

Range/Situation 1: Sniper and rifle range.
Range/Situation 2: Pistol range.
Range/Situation 3: Stand-off, showdown distance.

Range/Situation 4: Punching, kicking, stick and knife distances.
Range/Situation 5: Stand-up grappling.
Range/Situation 6: Standing vs. downed keeling (or seated).
Range/Situation 7: Kneeling vs kneeling.
Range/Situation 8: Ground topside.
Range/Distance 9: Ground bottom-side.
Range/Situation 10: Down on a side, a shoulder and hip down.

Rifle and/or maybe pistol range.   *Interview/argument, stand-off distances.*

Hitting, dueling, kick boxing distance.

Stand-up grappling distance.

Knee high vs. standing.

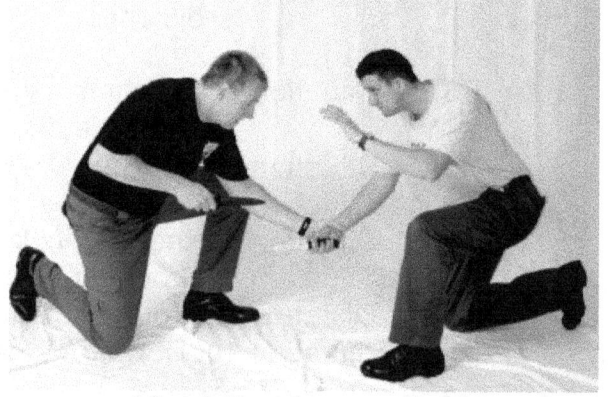
After falling, knee vs. knee.

Grounded vs. standing.

Grounded vs. kneeling.

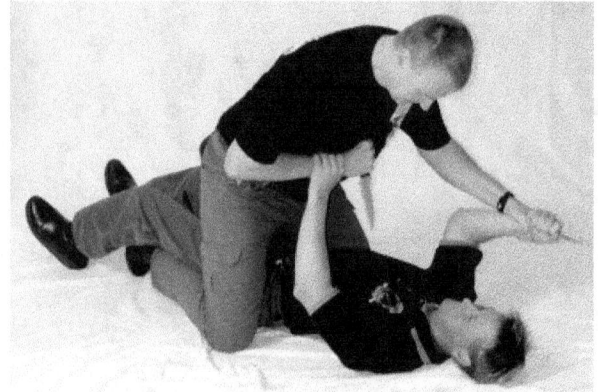
One topside, one bottom-side.

Side-by-side.

Universals - Getting Ready to Move

Before you move, you were "still" or relatively still. Citizens, police, security, even soldiers and Marines patrolling the streets of the world, become engaged in "face-to-face" encounters with people of all sorts, at all kinds of levels from introductions to explosive arguments. The stance is just before a solution, or departure, or hand, stick, knife, gun fight. How will you stand? Blade your body a bit? Not? But there are classic suggestions from veterans. Review them for sure, but you must select ones suitable for yourself and suitable for the situation. Each convey a subliminal suggestion. Also it should be noted that armed people like to keep their weapon "side" away from the problem person, protecting the weapon but also freeing up some "arm-space" to draw the weapon.

Totally relaxed.   *Conversational.*   *The thinker.*   *The prayer.*

Lapel holder.   *Hands off.*   *Torso twister.*   *It's on!*

The ambush has defeated the greatest militaries of the world. The element of surprise. How far and fast you move will depend on how alert you are just before you need to be. It seems like every time someone is victimized by a crime, or their borders invaded, a slew of "monday-morning-quarterbacks" declare that the victims were not alert enough. Even if the ambush was essentially unpredictable. In the most theorically sense, the complainers are correct.

There is a response time involved with an ambush attack or one you can see coming. This is not a book on how to stay alert and read potential danger signs, but since it involves the first instant you see fit to move a certain way in a fight from stand-up to the ground, it needs an introduction and mention here.

Standing. Seated. Kneeling and on the ground. Home. Away. Indoors. Outdoors. Where will you be? "Where" is just one big question and part of the *W's and H* survival questions. The best way to prepare for all attacks, ambush or otherwise is to use the *"Who, What, Where, When, How and Why"* questions as a basis for planning.

Who? Who are you and who do you really think will be attacking you? In terms of footwork and maneuvering, how fast and agile are you? Your endurance? His?

What? What do you really think will happen? What will you do? What will he do? In terms of footwork and maneuvering, what will you do? Move to defend? Move to escape. Move for a better position?

Where? Where will you be? Where is he? In terms of footwork and maneuvering, where can you move to? Are you in a snowstorm in Alaska? A beach in Jamaica? A phone booth? A parking lot? Do you know the possible escape routes?

When? When is this? Night time? Daytime? Winter? Summer? Holidays? In terms of footwork and maneuvering, when will you make your move? When will he?

How? How will you react? How we he? In terms of footwork and maneuvering, how will you handle this situation?

Why? Why are you still there? Why is he? In terms of footwork and maneuvering, why will you choose certain movements over others?

A sampling of answers. Where and when? The common denominator of attack for all parties is when standing, waiting, walking or perhaps jogging. Unless you are at the beach or park, or perhaps in bed at night, a citizen will not usually be attacked/ambushed while prone. A law enforcer may obviously be attacked while standing, but also while seated in a vehicle or while eating in a public place. A military person may be attacked while standing, running...while prone as a guard, lookout, defending the line, or even sleeping.

Why are you still there in a dangerous place and time? Remember that if you choose to just run away, you could be chased. Military and criminal history tells us that you are more easily killed from behind, as your face is turned away. The psychology, full intent (and fitness) of the attacker might be hard to predict. This is why the military suggests that you conduct an "orderly retreat" rather than just turning and running. A sample of orderly retreat might be pointing a weapon at an attacker and backing away for a distance, rather than just turning and running. No one can define an orderly retreat for all the circumstances and personalities of a situation.

But, follow this logic as you try to answer these *W and H Questions*. Obviously you need good intelligence info on the topics. And yes...try and stay alert to dangerous people in dangerous places at dangerous times.

You know that many survivors of airline crashes credit that "good shoes" helped them evacuate planes. When you think about it, "good" shoes would be handy in all tricky situations, wouldn't they? A footwork...universal!

An Interview Stance and Maneuvering There Within

So, if it's not a stone-cold, surprise ambush and you are in an encounter with someone picking a fight with you for whatever the reason, yelling and arguing with you, and things may go physically "south" very fast. How will you stand before this person? How much can you move around? In the best aspects of police work, even if someone is screaming at your face? It is still an "interview," and you have to somehow remain cool, calm and collect. This is professionalism at the highest police level, and maturity at the civilian level. Thus, the "interview stance" is a ready position, but not too ready as if to psychologically escalate the situation, or cause any possible witnesses to think YOU were escalating the situation into violence.

I would like to say that in all my decades of police work I had a trick in this pre-fight instance. If a situation was percolating into physical trouble, of course my body/feet would be bladed from that person, as in the right or left foot back from the other, but not too, too far. We call that "bladed." And I would, inside my pant's legs, unbeknownst to anyone, bend slightly at the knees. This was a more springboard, athletic position. This turned on the "juices" in my body that trouble was brewing.

Before I move away from this "spot," I want to bring up this body-bladed idea, because some people don't like it and offer all kinds of reasons for not. I don't agree with all those reasons and no matter how you stand, a Monday-Morning-Quarterback grappler has a way to take your down, or mess with you. Just get ready to move. A straight-on stance is one step forward or back to a bladed stance. Simple walking creates a bladed-hip position. It happens. When you move, it happens. The fight itself will change the push-pull of the footwork. Don't overthink this, it's situational and positional. Same with putting your weapon-carry-side back. It's situational and positional. Just be athletic.

But, if you jump into a fighting stance, "before" the actual fight, which might make good, pre-preemptive sense sometimes, be aware that your "action-guy" pose could be be perceived as an escalation to violence. This may spur on the other guy, or look like you're the actual fight-starter to ignorant witnesses.

"Two guys were just arguing, officer. And then, the guy in the suit squared off like he was going to fight, so the other guy did too. Then the fight started."

Part of everyone's repertoire should be the "preemptive" strike. You know this physical fight is going to happen, so you strike first. This is your own little ambush, usually delivered from a non-fighting stance. Could be an arm strike or a kick. Could be a carefully placed head butt?

That is why all unarmed, self defense systems must practice all their strikes and kicks from the typical non-fight stances, all which we will review later, along with the fighting ready performance stances/positions, moving and non-moving.

So, your pre-emptied STRIKE, though soundly sensible for your situation, could make you appear to be the instigator of the physical part of a fight. Needless to say, the same nearby witnesses may tattle,

"Two guys were just arguing, officer, and then the guy in the suit hauled off and smacked the other guy in the head!"

On the flip side of *your* knee bend is *his* knee bend. If you have a person confronting/standing before you, and he suddenly crouches down? This is a natural athletic move. This is NOT good for you. It is a positive indication that he is about to get physical. Look at the bent knees and the hands up in this picture in the upper left of this page. This is a fighting "stance" for when the fight starts. If they "spear" up their hands? Crouch and/or, start twisting their torso which is a high precentage, common precursor for a sucker punch, you must act according and prepare for trouble. Remember these cues when you have to explain yourself later.

Like so much in life, anything within a certain spectrum of events, good or bad, can happen. Things may work. May not work. In this particular "stand-off interview" moment, here are some common, even natural responses in the "script" of life.

    1: Leaving, fast or slowly.
    2: Cowering, "collapsing" of some facial and physical sort.
    3: Automatic anger.
    4: Ignoring, in some situations.
    5: Command presence of some sort, as in not collapsing.
    6: Fighting ready pose.

There are still options for maneuvering in a so-called stand-off confrontation, many are done with simple walking steps. You can:

    1: maneuver a distance from the problem person or persons.
    2: sometimes you can just keep walking.
    3: maneuver near an exit, for a sudden escape.
    4: maneuver over into the sight of witnesses or help.
    5: maneuver over to something that can be used as a weapon or a shield.
    6: realize a person or persons may be maneuvering *you* by crowding you, distracting you and getting you into a inescapable position, no-witness-around, situation. A pre-meditated ambush may be full of these factors.
    7: orderly retreat as previously discussed, or even running.

Quick Note on Verbal De-escalators

While on this initial confrontation subject, a quick "side step" here from physical movement. As for all the verbal, de-escalater experts out there, there are many courses available on de-escalation, run by lots of intellectual folks who have never had such attacks and confrontations forced on them, and over-value the idea that great, practiced orations will interrupt a fight.

You will hear advice from all sorts of people. Remember that deescalation for cops is different than for guards, for door men, for soldiers, for citizens on a parking lot, family members, or road rage encounters, etc. Run their advice though your "Ws and H" filter. While remaining within a spectrum of outcomes, the encounter is quite situational for you.

Never forget this guy drawn below. All your non-aggressive, micro-expressions and rehearsed non-aggressive wordings and steps won't stop him. He has his own script. He follows an anthesis to your script.

Universals - Moving Under, Around with Lights

Unless you're near the North or South Pole, about half the day on average is getting light and dark as the sun orbits. You need lights in the dark. Whether walking, jogging or

```
"Quit telling me to calm down!"
"Quit calling me, sir!"
"I am not your friend!"
"I am not your brother!"
"Yeah, talk civil, you pussy!"
"I am here to fight tonight!"
"I'll punch right through your palms."

  "You don't want any trouble?
   I came here for trouble!"
```

running around in the dark, moving atop the wild and whacky surfaces of the planet can be risky. Add into the mix, people trying to bust you up, rape, rob and, or kill you and matters get even worse. It could also be high noon on a bright and sunny day and a plumber, an electrician, a cop or janitor may step into the pitch black darkness of a basement.

The good news is rarely are you ever in complete blackness. There is almost always some level of ambient light glowing from somewhere unless you are in the center of a building or deep in a cave. Each person's optics vary, but medical experts report it may take a few minutes for your eyes to adjust to darkness. In an emergency you don't have that much time. Even with this adjustment, you are still handicapped.

The moral of these predicaments is - a person needs a flashlight or as some parts of the world call it, a torch. Quick light may be vital, quick light from handy equipment on scene like indoor and outdoor lighting, or equipment you carry - like flashlights.

If you have light, should you put it on? In a search for the suspect/enemy, whether you snap on a wall switch or click on a flashlight, should be a calculated event room by room. Yes, the light would be nice for a quick scan, but the enemy may also see you more clearly. You want to see them. They want to see you.

Light "looks both ways." A small, night-light in your kitchen, may make you feel all cozy and warm inside, but it allows a home invader to see the layout of your home from the a crack in a window curtain while he peeps in from outside. When an officer clicks on his flashlight, or draws his latest red laser beam, guided pistol, the criminal knows right where he stands.

A federal SWAT team once approached a house at night and when on the sidewalks, triggered the typical home motion-sensitive, outdoor security lights. The lights came on, the dangerous occupants were alerted, and the team was forced into a dynamic entry during which, one team member was killed! Now many teams carry suppressed .22 firearms to silently snipe out security lights, as well as run a pre-raid, recon for sensors.

Many combat vets report the flash of close gunfire at night disturbs their vision. Officers even complain that the spinning emergency lights on their squad cars can burn their eyes. Sudden light in the darkness has disabilities for you other than being spotted. It can blind you to some degree. This is why a flashlight beam into the eyes of a suspect you hold at bay is a solid strategy. If you can predict a sudden and temporary bright flash coming, such as a room light, or even a flare on the battlefield, try to keep one eye closed to maintain night vision in that one eye until the flash is over.

Today's flashlights are incredibly powerful for their size. They fit in ones hand like a martial arts, palm stick and the beam can be emitted from a saber grip or reverse grip. An unarmed person can walk or run with such a light in one hand with significant ease. The same for a person armed with a knife or stick.

A universal hand, stick, knife, pistol flashlight technique is to keep flashlight use to a minimum. A silent switch or push button can work in your favor for stealth.

Decades ago, trained professionals shot a pistol with one hand and two hands, and for most, did so 50-50. Today, very little time is spent shooting with one hand, maybe 85% two-hands, 15% one hand, even though statistics show that about half, or a little less than half of police and criminal shootings are close. Close enough to justify single-hand shooting, pistol somewhat back, and that thrusting your pistol out into a two-handed grip often places your gun too close to an opponent, into a lunge and reach of the suspect/enemy. Justification is not the best word. Smarter is a better word.

The old-school war weapon commandments go:

1: If thou art in a gun fight, thou shalt use a rifle.
2: If thou art does not have a rifle, thou shall use a pistol.
3: If thou art shoots a pistol, better to shoot it with two hands. (If far enough away from the enemy, if the support hand is free, if the support hand is uninjured, if....).
4: If thou cannot shoot with two hands, shoot with one hand.
5: Thou shall use your pistol to fight back to get your rifle.

Sounds good. When possible. I agree completely. These are pretty good commandments except, who among us, battles in the rifle range, rifle vs. rifle? Who among us carries a rifle around? Who has one always handy? Who among us is in a situation where we can use a pistol to do battle to get back to a rifle? And finally, since about half the gun fights are pretty darn close, who among us need shove our pistol out far front in a two-hand grip to have it slapped aside or grabbed, even disarmed? One handed grip, gun back is the solution.

Still, in our desperate attempt to score well on targets, to keep that steady hand, we predominately shoot with two hands like...like our lives depended on it, or at least bragging rights, bullseye scores. Yet, stats and common sense tell us often our lives depend on shooting with one hand. Still, a huge, mandatory, movement in two-handed, pistol shooting manifested as a be-all, end-all method, along with the mandatory accessing of sights.

After the "mandatory" two-hand acceptance, the flashlight industry invented these fabulous smaller flashlights, which confounded the target shooting industry - one mandating us to virtually shoot with two hands, ALL THE TIME! What in the world could they do? How can we hold these fabulous little flashlights and still shoot with two hands?

Probably the best solution for them was to mount the light right onto the pistol, and, or have lights built into the gun, even though bad guys ignore the "new rules book' and still instinctively shoot at the light. The proposed solution to that is to have a silent switch on the lights and leave the light off as much as possible. And of course, keep moving.

Minus the attached pistol light, people invented some supported configurations, desperate to keep as close to the two-hand-grip concept as possible. Various "positions" were invented and the inventors were blessed with having their configurations named after them! Like they were doctors with cures for cancer or some such thing. Pistol experts still whisper that these are not good substitutes for a two-hand grip, as the support is rather lame.

Violating the too-close-to-the-body rule, yet still popular, here are three very common flashlight-supporting, pistol-positions, 9, 3 and 6 o'clock. People like to have silent on-and-off switches and buttons and try to keep the light on at a minimum. You walk/search while holding the light. By the way, gun people argue a lot about the validity of these methods.

Where will the bad guy shoot at you now? But this method does light up your pistol sights.

Light works both ways. You are spotted.

Stay out of the classic "Fatal Funnels" as much and as fast as possible.

In over three decades of military and police work, private investigations, security and bodyguard assignments, I have searched probably a thousand or more homes, businesses, buildings, fields, arenas, alleys, military installations, foreign villages, etc., and I refuse to limit myself to any one stance or any one flashlight-holding method. Here is what I and many veteran experts feel are important about searching with hand-held lights and in darkness.

1: Carry a light with a silent on-and-off switch/button/pressure pad.
2: Be flexible in how you hold your light in conjunction with your pistol.
3: Your long gun probably needs an attached light with on and off silent touch pads/switches.
4: Limit your exposure by limiting the use of your light.
5: Gas, fog, smog and smoke can interfere with your flashlight beam.
6: Maintain night vision by closing one eye if suddenly faced with temporary bright light.
7: Be aware of ambient light in the environment that may give your position away. There is, almost always, some sort of ambient light.
8: Use your light (if very powerful) to distract and stun the eyes of your suspect if you catch him.
9: Using a light and handling and shooting a pistol is a proven coordination issue, especially when you cross your wrists. Train for it.
10: You do not have to aim light like a "sight picture" on like a pistol. Use your gun for that chore.
11: Searching in the dark? Do not look directly at something. A general stare into a large area is more likely to catch unusual movement in dark environments. An old military adage is to look, at the "sides of something," rather than right at its center.

12: Searching outdoors in the dark? Look at the landscape and let your brain absorb the natural movement, such as that of the wind. Look at the big picture. Your eyes will often zero in on the odd, abnormal moves in a manner that doesn't fit the natural, overall flow.
13: You know you and your light will be seen. Try to maneuver in such a way as to limit your specific location. The suspect/enemy may have an idea where you are, but not exactly where you are. Turn light on and off lights and keep moving.
14: Crawl, walk, jog, run while holding a flashlight when training.

Note: More on searching/moving techniques later in this book. This was just an introduction to the universal use of lights while moving in general.

Universals - Crawl, Walk, Jog, Run

"If you can't fly then run, if you can't run then walk, if you can't walk then crawl, but whatever you do you have to keep moving forward." – Martin Luther King, Jr

And the last universal, is making the connection between everyday movement. We started out crawling although most people do not crawl around much, if at all, and even if your job requires you to, it is not as often as being upright. We start out crawling, and graduate to walking, then jogging and for some? Running.

Most martial art systems take it upon themselves to virtually teach people how to walk all over again. They "baby-step" you through various, primitive walking movements and if honest, if thoughtful, their end game is make you walk, run and turn in all directions to fight. I am not sure most folks realize it, though. Then you have to do these moves "while holding," as we call it, holding sticks, knives and guns.

Fighting footwork certainly also resembles sport footwork and maneuvering. But I don't believe you have to learn to walk all over again to develop fighting footwork.

I think that the word "warrior" is way over-used and over-done. But I do use it in one sense, "warriors run." Warriors must cover ground, and be as fleet of foot as they can be. They must advance and retreat. Never stop your running training. It should include sprinting.

Universals - The Stress Weapon Quick Draw While Moving

Another hand, stick, knife, gun universal is drawing weapons while standing still and certainly when moving. And, in the case of long guns, getting them "up" or off your shoulder and into an active position. Study these specialized subjects. Know you have to do them on the move. I will offer this overview here. Your pistol, stick, expandable baton or knife will be secured in 3 sites:

   1: Primary Carry Site: This is your belt line and your pockets. Think "quick draw."
   2: Secondary Carry Site: This is a bit deeper, under the shirt like a neck knife, or inside your boot. Buried a bit. Think "back-up."
   3: Tertiary Carry Site: This is a weapon stored off the body. Think "lunge and reach."

Pick your carry sites. Now, how do you reach for that weapon? Can you do it...
   - standing still?   - while walking?
   - while running?   - while on the ground? (Top? Bottom? Right side? Left side?)

Drawing while walking.   *Drawing while running.*   *Drawing while deceiving.*

Study the hand, stick, knife, gun "draws" in my other books, films, seminars and the works of others. Okay, now back on the footwork and maneuvering track!

Universals - Body Elasticity/Body Maneuvering Before and Later with Footwork

Hoping that footwork is the only tool needed for escaping strikes when standing, or only using big maneuvers to escape strikes on the ground, is just not comprehensive. Maneuvering your body without footwork or changing big body positions on the ground such as in ducking, dodging, twisting, leaning, etc. are important first steps.

Many attacks are sudden, ambush or ambush-like, and are so fast that people cannot step or run off, so body elasticity is very important in those first seconds. It is also important to support blocking and counter-attacking. Many people are not "loose" enough to do this. This must be developed in training. One can work on specific drills which I will offer here, or other exercises that help flexibility (even Yoga. It is wiser to work on goal-specific, goals. Yoga is abstract, while slipping a punch is...slipping a punch.)

A combination of body elasticity, blocking and footwork/maneuvering is optimizing ones safety. A classic, accepted method is to approach these skill developments separately and end with all three together. One drill I teach is called the *Dodge/Evasion Drill* for starters. Then I add counter-attack options and the title of the drill gets longer to *Dodge/Evasion/Counter Drill*. Since this book is dedicated only to footwork and maneuvering, we will not be doing any blocking or counter-attacking here.

The first part of the drill has NO FOOTWORK, and NO BLOCKING so as to isolate, develop and maximize body elasticity alone. After we add footwork exercises and ground maneuvering in the upcoming chapters, this drill format will reappear again later. But the big picture contains many elements.

The Dodge/Evasion Exercises

This drill was inspired by a Filipino martial arts set. The original master set was too simple for me, so I added more strikes. These additions allowed the drill to apply hand, stick and knife attacks.

These are the 10 attacks. They develop the elastic skills of body evasion. A trainer starts the attacks with a stick set, then attacks with a knife set (see below), then attacks with hand strikes and kicks. The trainee evades with the suggested moves. There is no footwork at first. Footwork escapes would diminish the need for body elasticity and evasion. Initially the trainee's arms also dodge the attack. Later in the skill progression, the arms will block, footwork will be added, etc.

First we begin with the 10 common attacks:
Attacker Set 1: Two high hooking attacks.
Attacker Set 2: Two middle height (stomach) hooking attacks.
Attacker Set 3: Two low (thigh/knee) hooking attacks.
Attacker Set 4: Two vertical center line hooking attacks.
Attacker Set 5: Two stabs, one to stomach area, one to the head and neck area.

Attack with a knife.
Attack with a stick.
Attack with punches and kicks.

Ambush attack 1: Head shot from trainer's right. An inward strike.

Ambush attack 2: Head shot from trainer's left. A backhand strike.

Ambush attack 3: Belly shot from trainer's right. An inward strike.

Ambush attack 4: Belly shot from trainer's left. A backhand strike.

Ambush attack 5: Knee shot from trainer's right. An inward strike.

Ambush attack 6: Knee shot from trainer's left. A backhand strike.

Ambush attack 7: Shot up from trainer's low. Groin target? An upward strike.

Ambush attack 8: Head or clavicle shot from trainer's high. A downward strike.

From any ambush start...

Ambush attack 9: Belly stab.

Ambush attack 10: Face stab.

*The trainee uses body elasticity without footwork to dodge the attacks.
(On the knee attacks, we expect a little escape footwork.)*

Attack on the ground. The trainer attacks on all these 10 angles

Defender vs. the Standing Attacker
The defender will quickly realize that the standing movements don't equate well with being on the ground. He or she must improvise to get out of the way. They will:

- Head shift side to side.
- Torso shift side to side.
- Body roll.
- Spin.
- Leg left.
- Etc.

Note: At the end of each one, or the set of ten - the trainee pops up to a standing position.

Defend while held. The trainer attacks on all these 10 angles

This causes the trainee to REALLY use his or her body to the max to dodge the attacks.

The Dodge/Evasion series continues, we will add the following. Footwork is next!

- Blocking.
- Footwork.
- Hitting and kicking back.
- Unarmed vs. weapons.
- Knife vs. knife.
- Knife vs. stick, stick vs. knife, stick vs. stick.
- Finishing with takedowns.

Universals - Running and Sprinting

The definition of a sprint is to run as fast as you can for shorter distances than a long run, or as fast as you can for as long as you can.

Other than the usual treadmill advice, Jon-Erik Kawamoto, CSCS, CEP of *Bodybuilding.com* offers other inside-the-gym advise to get faster. "Incorporate a progressive strength-training program that focuses on exercises such as deep squats, deadlifts, power cleans, and lunges. Don't worry about being fancy. These recommended exercises transfer over to the tarmac and improve your ability to recruit the fast-twitch muscle fibers necessary for powerful locomotion. I recommend implementing these exercises into your program 2-3 times a week to improve your relative and absolute strength levels. Explosive plyometric training like squat jumps, hurdle hops, and bounding has been shown to increase speed by shortening ground contact times and increasing stride frequency."

Donovin Darius for *NFL.com* suggests, "Nothing helps you learn how to run faster than running itself. Now that you have the information it takes to improve your form, it is not time to put it all together. For this exercise go to a field or surface where you can run for 30 to 50 yards straight. Your focus is running with perfect form: Hands moving "cheek to cheek," knees coming up waist high with quick and powerful steps that cover a lot of yards with each stride. Perform this exercise by marking off 30-50 yards and on command sprint from start to finish. Remember to get a good warm up and stretch before performing this all-out sprint. Start out running five sprints with three minutes of rest in between reps. Add one more sprint each week until you can perform 10 in one practice session."

One controversial point I discovered about this sprint training is the argument about step counting. Numerous coaches tell you to shorten your steps to be faster. Numerous other coaches suggest you lengthen your steps to cover more distance in less time.

Citizens, soldiers/Marines and enforcers must remember that "combat sprinting" is different than sport sprinting. We are not running in a specific distance for a specific time in skimpy shorts, tanks tops and, for the ladies-sports, bras. We are running over unpredictable distances in street clothes, street shoes, carrying gear and for some, what I call "whole holding," as in carrying various weapons. Still, I want you to refer to expert sport coaches about the specifics of sprinting and experiment with what might apply to crime-survival, crime-fighting and war.

Need we review here the reasons why citizens, soldiers/Marines, enforcers and sport athletes know they need to run for distances longer than sprints? Sprinting in spurts is one thing. "Running" is something else. Most people hate to run. I know bodybuilders that, fearing even a 1/16th reduction in their precious muscle mass, will shun running (or sprinting). They quote muscle magazines that say running is not good for you. But, warriors run. They cover ground. They jog; they dash; they hop; they leap; they zig zag; they move through space.

Never stop running as long as your legs still work. Run in all kinds of weather. A warrior toughens his or her soul by experiencing that discomfort that comes from running. The residual benefits are also vital. Run while holding a training knife, a gun, or any tool you might carry!

Can you name a successful athlete who didn't or doesn't run? There are even champion chess players that put in the roadwork.

In my own observations in athletics and survival over the decades, I think there are two kinds of running postures that relate to fighting. A speed run and a combat run. In a speed run you can remain upright, chest out, as suggested by all racing coaches. In a combat run, fearing to be seen or shot at, you have to run bent over. You might sacrifice some speed, but prolong your life.

Two kinds of running:
* Speed run. You remain upright, chest out, as seen by track sprinters.

* Combat run. Fearing to be seen or shot at, you have to run bent over. You might sacrifice some speed, but prolong your life.

"The fight is won or lost far away from witnesses, behind the lines, in the gym, and out there on the road, long before I dance under those lights."

Muhammad Ali

While you are running:
* practice drawing weapons.
* run while holding weapons.
* run while holding mission-needed gear.
* run while wearing mission clothing.
* run on challenging surfaces.
* strike and kick while running.

Segment 2: The Stand-Up Footwork

Combat footwork is a mixture between sport footwork as in boxing and kick boxing. A practitioner can be unarmed, holding a firearm, a stick or a knife, or, any weapon. Here are the basic patterns, as expressed through the Combat Clock format.

First off, we must recognize that the clock, laid flat on the ground, has 12 numbers or 12 directions. And, the clock also has an axis point in the center.

You will pass through this axis point in some of this footwork, or one foot will be standing on the axis. So there are 12 numbers and an axis point to consider.

For basic training, we can refer to 12, 3, 6 and 9. Twelve being anything forward, 3 being anything to the right, 6 being anything backward, 9 being anything to the left. The advanced clock delineates even more.

We will explore 12 Stand-Up Footwork Concepts and Exercises. They are:

Routine Number 1: "The Ready Stance" to the "Get Ready-Bladed" Stance.

Routine Number 2: The In and Out, Stationary Axis Footwork.

Routine Number 3: The V Footwork.

Routine Number 4: The Upside Down V Footwork.

Routine Number 5: The Thai Jump Leg Switch.

Routine Number 6: The Shuffle (the Pendulum) Footwork.

Routine Number 7: The Step and Slide Footwork.

Routine Number 8: The Quarter, the Half, or the Full Circle Footwork.

Routine Number 9: The Pivot – Lead Leg or Rear Leg Footwork.

Routine Number 10: Walking - Walking off the Clock Footwork.

Routine Number 11: Running - Running off the Clock Footwork.

Routine Number 12: All Combinations Like "Half a V with a side-step."

Routine Number 1: The Ready Stance or "Step to Get Ready." The first step. Many beginners may not know this, but many beginning moves in a confrontation or weapon draw, usually involve taking a step forward or a step backward. Often a questionable and suspicious encounter or interview requires a person to "blade their body" somewhat sideways, accomplished with a single step.

The right foot steps forward or back, or the left foot steps forward of back. Which one? Pick the smartest position for the situation. People armed with right side weapons often want their right side back, keeping the weapon away from the enemy/suspect. Plus, it's better to draw your weapon from the rear side. Less chance for a countering problem. Most pros call it "blading" away from your enemy.

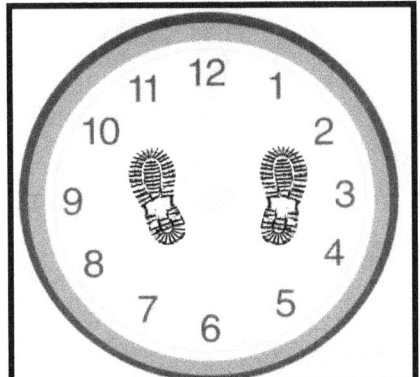 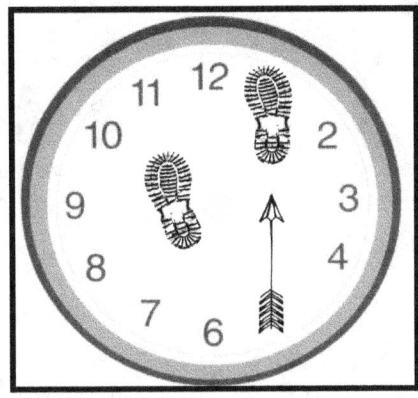 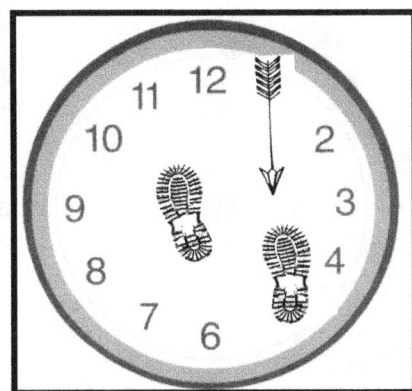

A common straight stance.   Right foot steps forward, or...   Right foot steps back.

The bladed, slightly stance is usually taught/achieved with these 4 options:
- Option 1: The right foot steps forward, or
- Option 2: The right foot steps back, or
- Option 3: The left foot steps forward, or
- Option 4: The left foot steps back

People armed often chose to keep their weapon-carry-side back. This is good for retention as well as creating space for the weapon-side limb to draw the weapon, away and as unencumbered as possible from the problem person's interruption.

A sample of this movement with a right foot step. You start out straight on to a problem. It escalates a little. You "blade off" to more of a side angle for a possible athletic response. Here, the right foot goes forward or backward as needed.

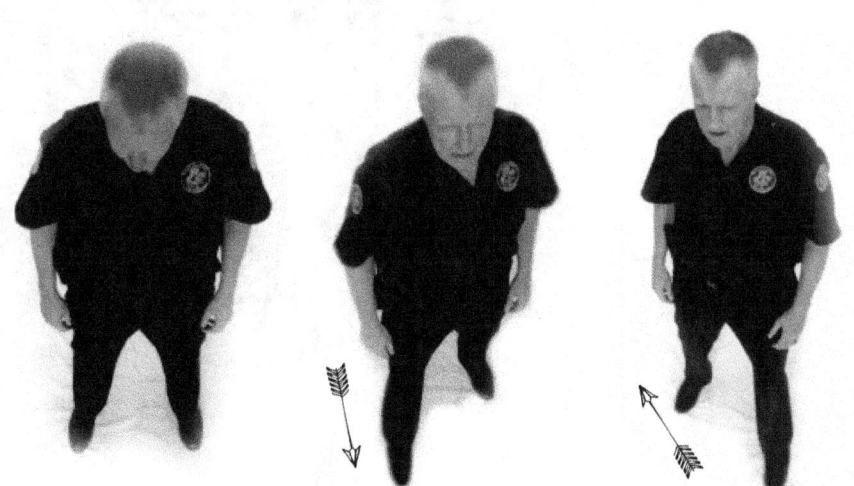

A distant, straight on stance. With 1-step, it changes. *Here, with a concealed, saber grip stick.* *Here, a concealed reverse grip stick.* *One step can change everything.*

The single right or left step, forward or back, right or left foot, to a more ready position is done, and should be practiced:

- Unarmed.
- With an undrawn pistol, or knife, or stick/baton.
- With a shouldered or lanyard long gun.
- With a drawn pistol, knife, or stick/baton.
- With a dismounted and, or up and ready long gun.

Routine Number 2: In and Out. The Stationary Axis. Either the left or right foot remains on the center axis while the other foot travels in and out. This keeps the fighter in closer. He moves out a bit, then back in. Right foot in and out, evading a hand, stick, or knife attack. The left foot remains in the center. The lead right foot moves from about 2 o'clock to about 5 o'clock and then back to 2 o'clock again. Rather then shuffle away and shuffle back, this keeps you closer. Switch sides.

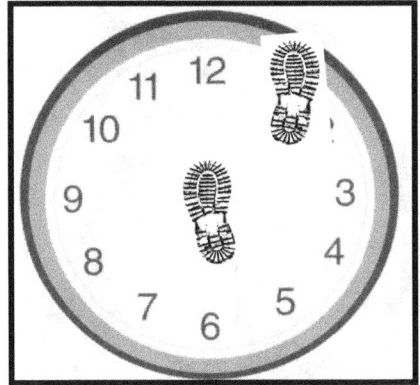 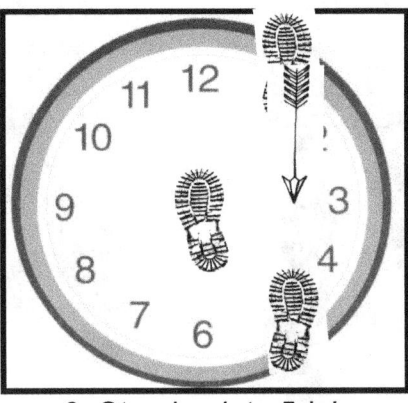 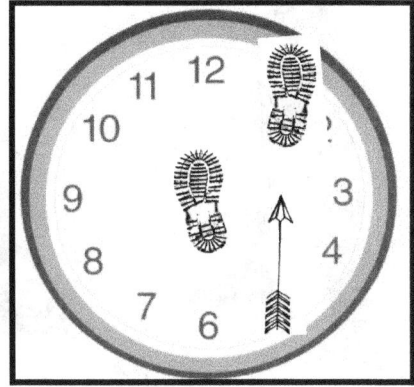

1: Right foot lead start.   2: Step back to 5-ish.   3: Step forward to 2-ish.

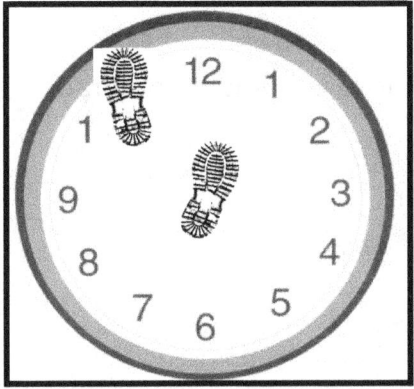 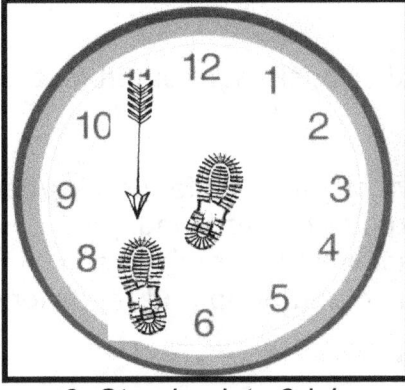 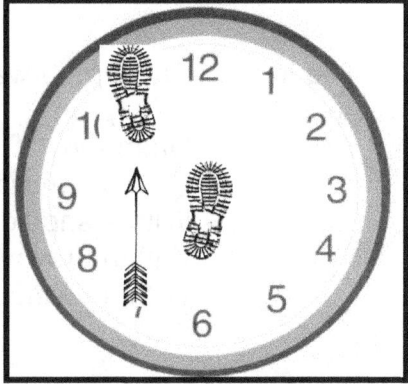

1: Left foot lead start.   2: Step back to 8-ish.   3: Step forward to 10-ish.

Here is a sample of a right foot in-and-out step. The left foot remains. You can step out and right back in. You do not lose ground because the left foot remains. Obviously do the same with the left foot.

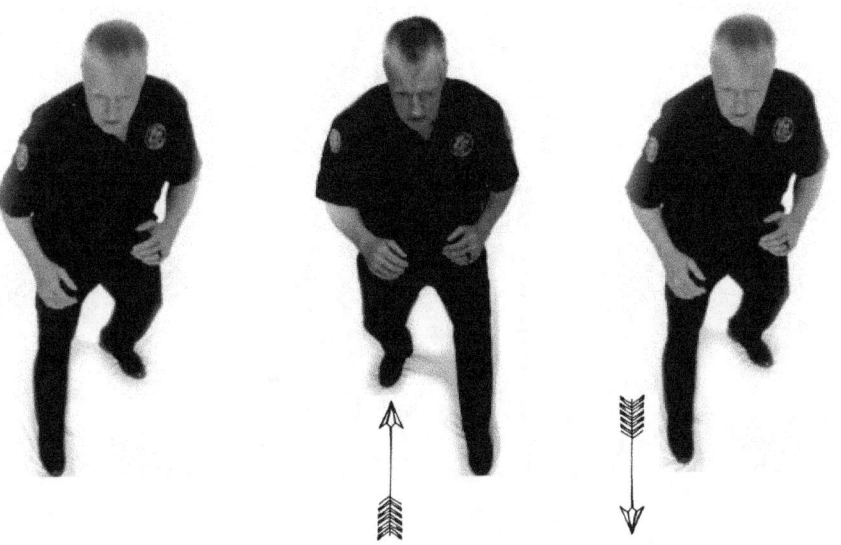

 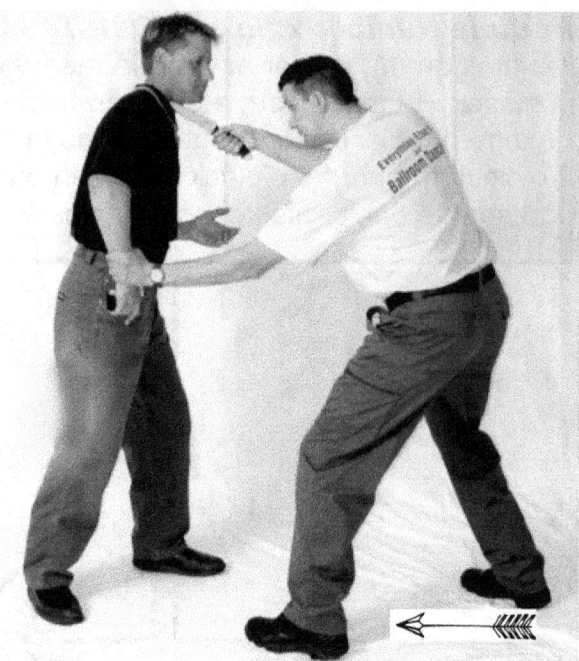

This in-and-out step keeps you close to attack or counter-attack.

The in and out step, forward or back, right or left foot is done, and should be practiced:
- Unarmed
- With an undrawn pistol, or knife, or stick/baton.
- With a shouldered or lanyard long gun.
- With a drawn pistol, knife, or stick/baton.
- With a dismounted and/or up and ready long gun.

Routine Number 3: The Letter V Footwork. This is another classic pattern. You start out with a right lead with the right foot at about 2 o'clock and the left foot on or about the axis. The right foot comes back to the axis as the left foot shoots out to, on or about 10 o'clock. Then reverse that pattern for an exercise routine. This can put you on the outside of an opponent's thrusting attack. You, of course, could start with the left lead and "V" off to the right. Step back and continue the V pattern.

Follow the "V."

This sample, right foot comes to the clock axis.

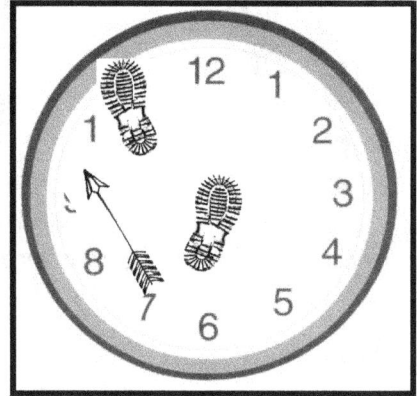

Left foot steps to 10-ish.

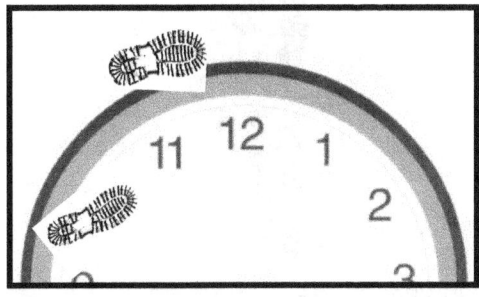

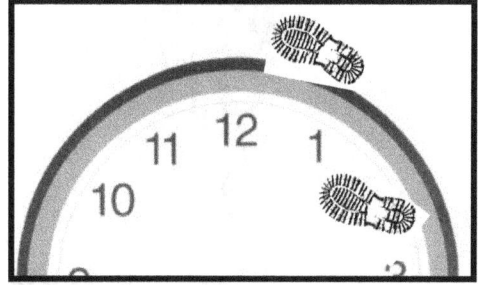

Pivot in at the top.
Do the "V' for a series. Then, start a series where, at the top of the "V", at the 2 and 10 o' clock points, pivot/turn inwards. This is when you get to the outside of an opponent and turn in toward them to engage them. (More on pivoting later.)

In this sample to the right, one starts with a lead foot, then brings that lead foot in - don't spend time there! You can't. Then shoot the other foot out in a lead changer. Practitioners work this V back and forth for exercise. This is often called "Triangular Footwork."

To complete the classic triangle, some practitioners finish with a side-to-side step and then a back step to return to the position in first photo. This completes an official triangle.

Also, it is important to practice pivoting inward at the 2 and 10 o'clock positions to turn in and attack.

The classic V Footwork is often followed up with:
    1) A pivot inward at the 2 and 10 o'clock, top positions.
    2) A side step to finish off the triangle.

Work these patterns:
- Unarmed.
- With an undrawn pistol, or knife, or stick/baton.
- With a shouldered or lanyard long gun.
- With a drawn pistol, knife, or stick/baton.
- With a dismounted and, or up and ready long gun.

Routine Number 4: The Upside Down Letter V Footwork. This is another classic pattern, often included in the topic of "triangle footwork." You start out, say, with a left lead with that left foot on the axis and the right foot at about 4 o'clock. The right foot comes to the axis and the left foot shoots back to about 8 o'clock. Then reverse that pattern/path for an exercise routine.

Follow the upside down V.

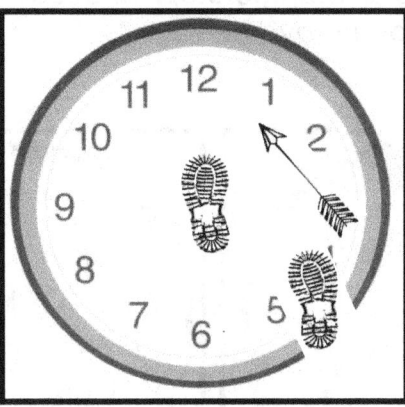

1: Starting position sample.

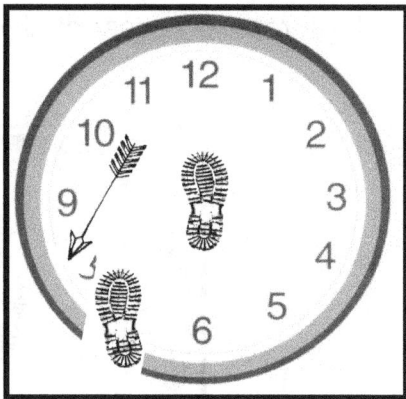

2: Right displaces left. Left shoots back to 8 o'clock.

The classic Upside Down V Footwork is sometimes followed up with a side step to finish off the official triangle.

Work these patterns:
- Unarmed.
- With an undrawn pistol, or knife, or stick/baton.
- With a shouldered or lanyard long gun.
- With a drawn pistol, knife, or stick/baton.
- With a dismounted and, or up and ready long gun.

Routine Number 5: The Thai Jump Leg Switch. This piece of footwork "hops" or jumps from right lead to left lead, and vice versa. For example, the classic Thai boxer does not wish to kick with a lead leg, only the rear leg. It is Thai doctrine that the rear leg be used. So if one wishes to kick from the lead leg side, to gain a powerful kick they jump/switch leads and deliver the kick from the rear. We can use this simple method to remove a "triangle step" and just switch leads in a "hop, skip and a jump." It is essentially one movement.

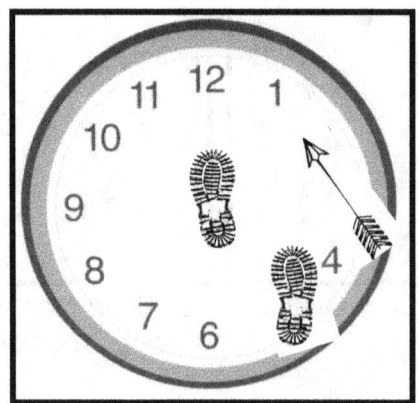 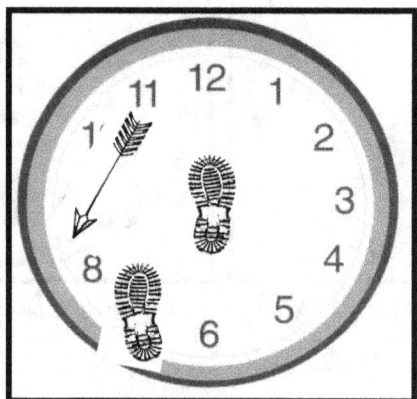

*You jump from one lead to the othe lead.
The "jumping" feet are barely off the ground.*

Work this movement:
- Unarmed.
- With an undrawn pistol, or knife, or stick/baton.
- With a shouldered or lanyard long gun.
- With a drawn pistol, knife, or stick/baton.
- With a dismounted and, or up and ready long gun.

Routine Number 6: Shuffle Footwork (the Pendulum). Think of a pendulum. One foot moves up to the other foot. Then that foot moves on to your selected direction. Your feet do not have to actually hit together. Your rear foot does not have to "knock" the front one forward as you might think if you fully define the word "displacement." But this is sometimes a wise practice for the beginner just to learn the concept. This is an exceptionally good move when increasing or decreasing the gap between you and your opponent in a retreat and/or an advance. The whole clock can be worked in this manner. Here are three samples.

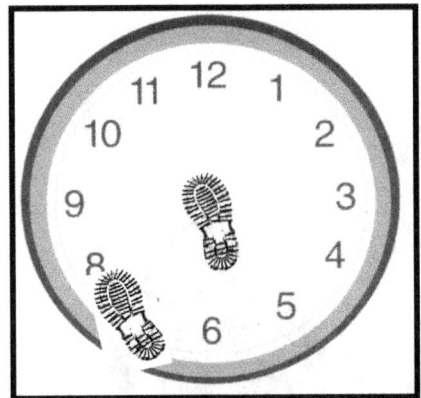
1: About a 7 to 2 shuffle.

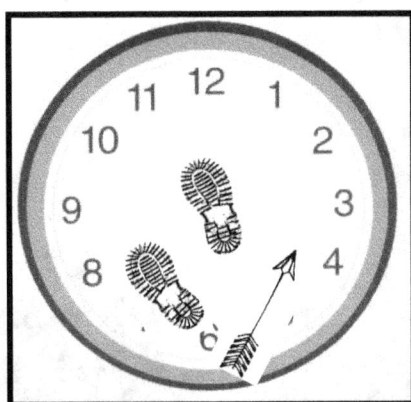
2: Your rear foot moves up.

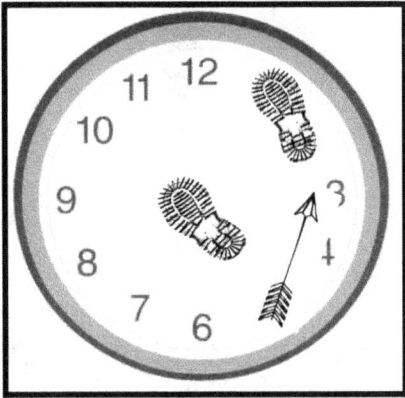
3: Your lead foot moves up.

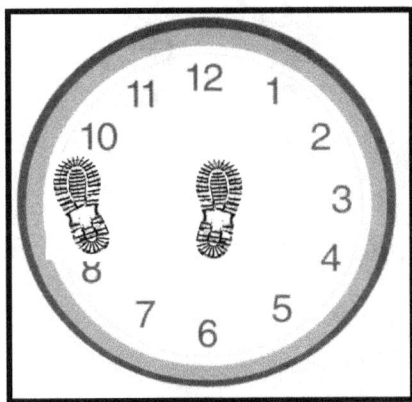
1: About a 9 to 3 shuffle.

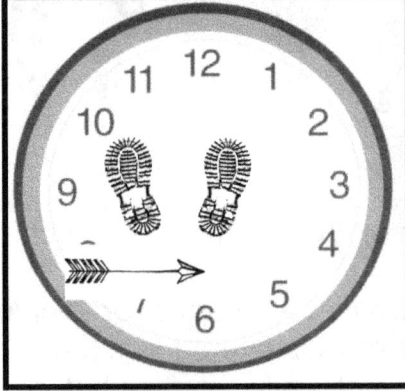
2: Your left foot moves up.

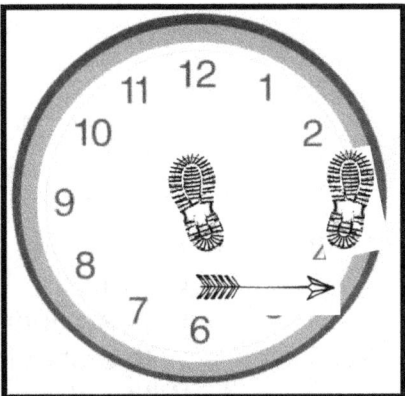
3: Your right foot moves up.

1: About a 11 to 4 shuffle.

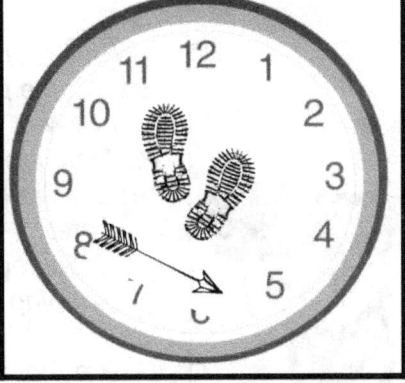
2: Your left foot moves back.

3: Your right foot moves back.

In this 10 o'clock to 5 o'clock retreat example, the lead left foot pendulums back to the center and the right rear foot goes back.

Here is a side-to-side, lateral movement. This sample has a shift to his right. His left foot shifts over. His right foot shifts over.

Work these movements on all the numbers of the clock, forward and back:
- Unarmed.
- With an undrawn pistol, or knife, or stick/baton.
- With a shouldered or lanyard long gun.
- With a drawn pistol, knife, or stick/baton.
- With a dismounted and, or up and ready long gun.

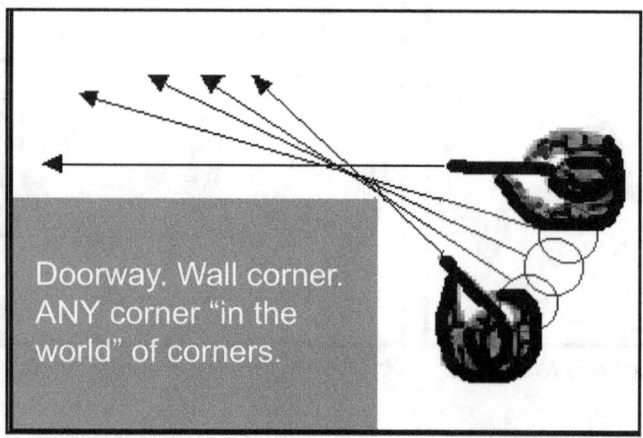

Doorway. Wall corner. ANY corner "in the world" of corners.

Side step footwork is frequently used in "slicing the pie." Each side-step opens more of a view of what's "around the corner." Professionals choose to:
1: Step and quick peek, jerk back for safety.
2: Step, look and stay, clearing and keeping that space you see.

Probably keep both in mind for situational problem-solving. More on this later.

Routine Number 7: Step and Slide. In Routine Number 5 the rear leg shot forward first, *displacing* the front foot. With this exercise, the lead foot steps off first and the rear foot slides up, follows up, to a body-balanced position behind that lead step. This can be executed forward and back and side-to-side positions. The whole clock. Here are two clock sample pieces of footwork.

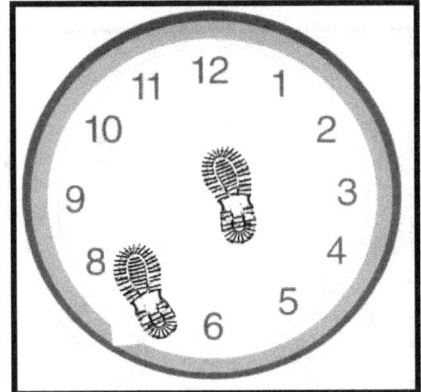
1: The ready stance.

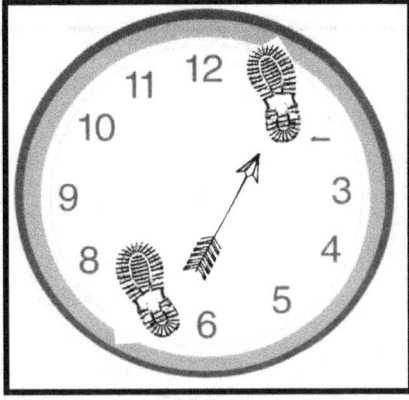
2: The right foot steps forward.

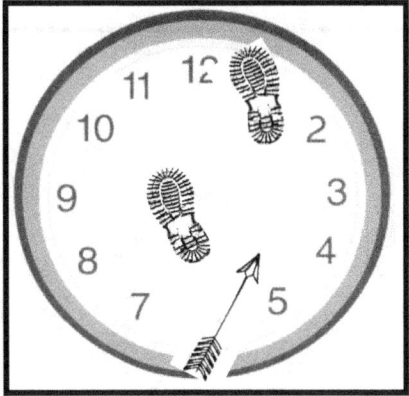
3: The left foot slides forward.

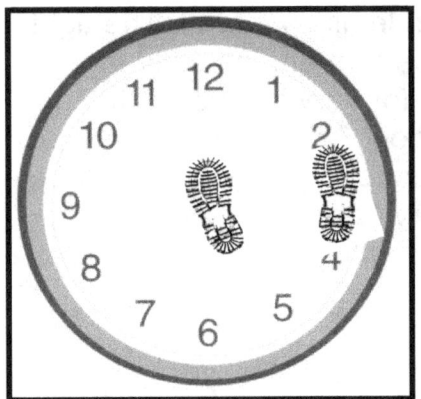
1: The starting stance.

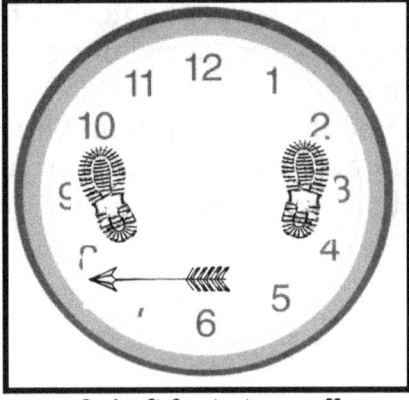
2: Left foot steps off.

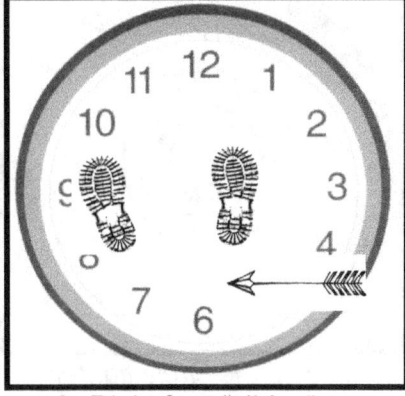
3: Right foot "slides" up.

Routine Number 8: The Quarter, the Half or the Full Circle Moves. Many boxers and kick boxers "hop/bounce" off in round patterns with their chests facing the axis, center of the circle. Sometimes they just need a few numbers to travel, sometimes more. The full circle is just an exercise pattern. Go clockwise. Go counter-clockwise.

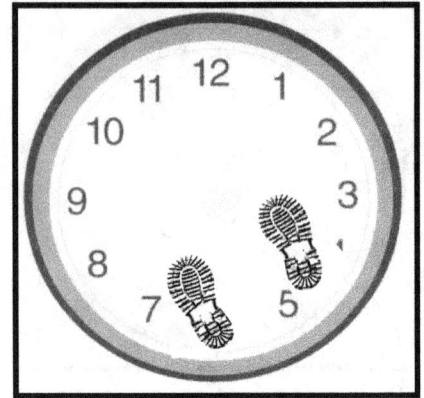 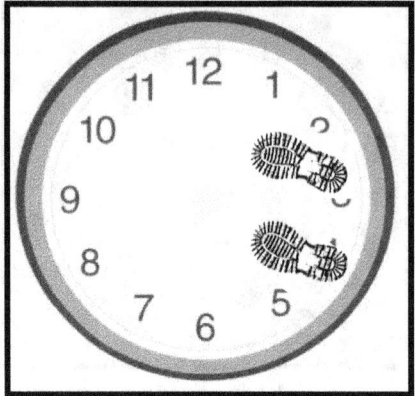 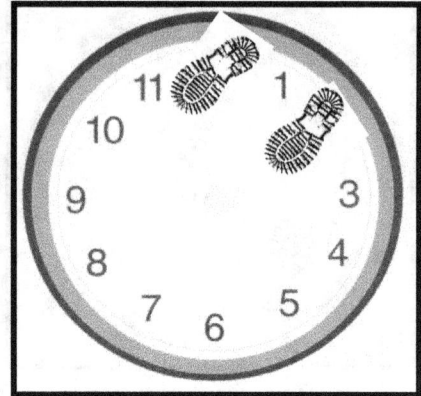

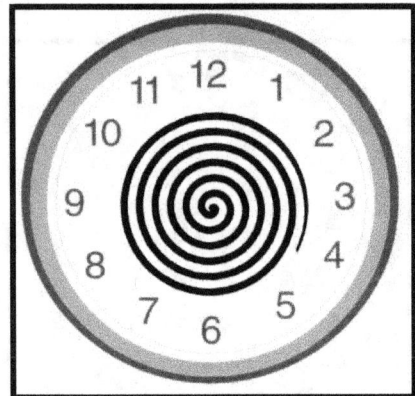

Closing the circle tighter.

You also hop, facing inward, in a circle. You can also hop in an ever tightening circle, like a spiral. The tighter you go, the more concise the turn and actually the more practical the skill gets for very in close fighting. Hop the spiral in both directions. In and out. Hand up, or dealing with whatever weapon-position/barrel direction you chose. This is a very important exercise.

Work these movements on all the numbers of the clock, forward and back:
- Unarmed.
- With an undrawn pistol, or knife, or stick/baton.
- With a shouldered or lanyard long gun.
- With a drawn pistol, knife, or stick/baton.
- With a dismounted and, or up and ready long gun.

Routine Number 9: The Pivot - Lead Leg, or Rear Leg. Clockwise and counter-clockwise pivoting. Using one foot as a pivot point (the figurative clock axis center, use your other foot to push yourself around clockwise. Or counter-clockwise. The center foot's heel raises and you pivot on the ball of your foot. You do this in every-day life when make a sharp turn, like in a supermarket or in your kitchen.

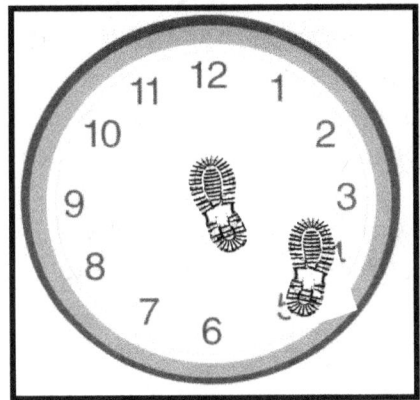 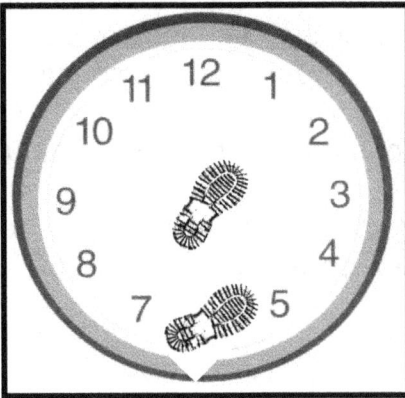 

1: You are ready. Raise your center foot heel.   2: Your other foot moves back.   3: Your other foot moves back. You turn with the step.

Work these movements on all the numbers of the clock, forward and back:
- Unarmed.
- With an undrawn pistol, or knife, or stick/baton.
- With a shouldered or lanyard long gun.
- With a drawn pistol, knife, or stick/baton.
- With a dismounted and/or up and ready long gun.
    Note: Remember the pivot inward at the top of the V footwork. Very important.

Routine Number 10: Walking off the clock with and without Firearms. The trainer creates lanes. The trainee must walk them quietly without firearms. Then, with a pistol and a long gun, the trainee must also walk them quietly and with barrel control. Where possible in the case of firearms, the trainee might shoot as he or she progresses to certain landmarks or sees cues. Turns are required training for the lanes.

Military experts say that if you walk with your legs far apart, this may cause a tendency toward side-to-side "barrel sway." If knees are closer the weapon may go up and down a bit. Awareness of this helps. Stride length counts inside this too.

Most experts suggest a slightly lower, what they call, "Groucho Marx" walk, as in a lower center of gravity. Going forward they suggest "rolling" your feet, heel to toe. Going backward it's toe to heel.

This type of "careful-step" training tends to slow a trainee down. There are many times when you cannot search or escape slowly and you just have to move fast or run.

Routine Number 11: Sprint Off the Clock. In this exercise, a trainee stands in a relaxed, hands-down stance at the center of the clock. A trainer calls out a clock number and the trainee must sprint off to that number. Running off between 10 o'clock through 12 to 2 o'clock is simple. Crouch and explode. But 3 thru 6 to 9 o'clock requires a turn. I suggest you do not pivot on your foot and run, but rather "spring-jump" off into that direction to create explosive power.

Work these movements on all the numbers of the clock:
- Unarmed.
- With an undrawn pistol, or knife, or stick/baton.
- With a shouldered or lanyard long gun.
- With a drawn pistol, knife, or stick/baton.
- With a dismounted and, or up and ready long gun.

Routine Number 12: Any Combinations Like "Half a V with a side-Step." Combination routines. We have already listed some when we used the side-step to complete the V footwork patterns. Each one of these footwork routines can be joined for advanced exercises, such as with this example. Once you begin combining these routines you have quite a list to work on. The list of combinations is lengthy. Here is one example.

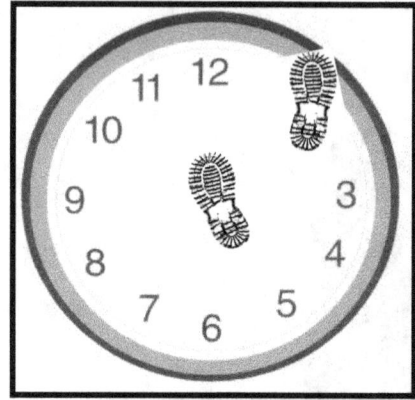
1: The ready position.

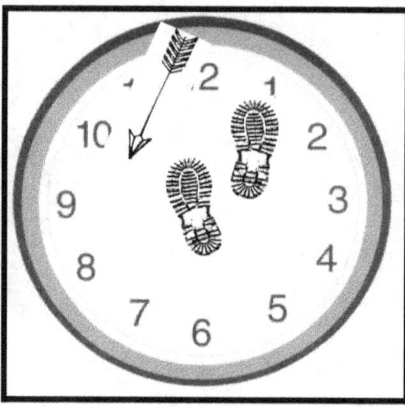
2: The right foot "Vs" back.

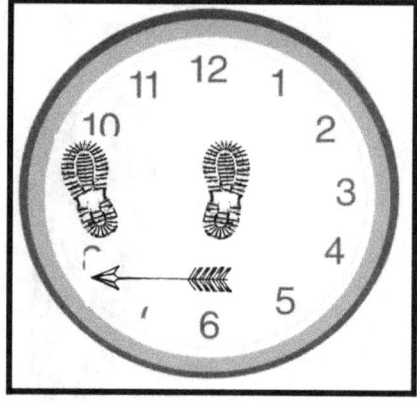
3: The left foot side-steps.

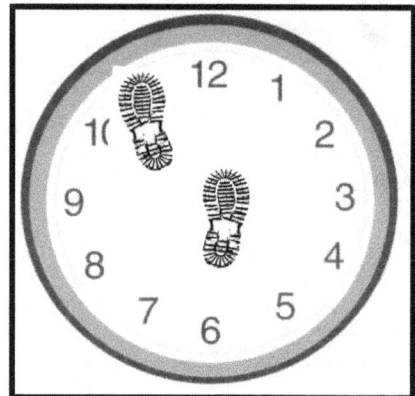
1: The ready position.

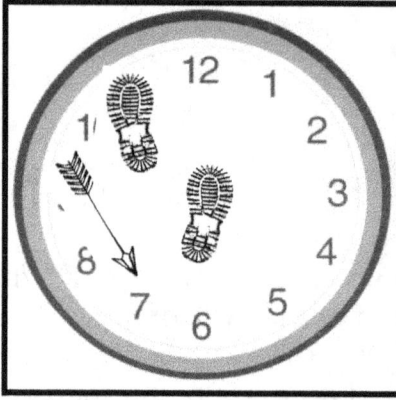
2: The left foot "Vs" back.

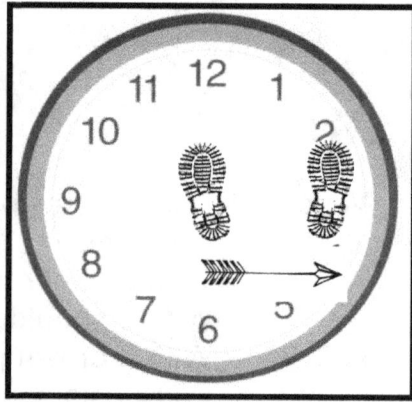
3: The right foot side-steps.

Notes on Footwork

These standing movements position you for boxing-like encounters or grappling.

1: Movements like step and turn in are also important in grappling.

2: A pivot, turn-in, and...

3: The other arm wraps, turn and run off! Imagine the damage.

Kicking Footwork

So far these samples of footwork also apply to kicking skills. They set up the momentum and positioning, the synergy for standing kicks. But, this is not a book on how to kick, so we will only worry about footwork here. Here are but a few samples.

Here is a sample of some pendulum-like footwork, which also helps close the gap.

Here is a closer range kick sample that has no real footwork. The classic "Cat Stance." You put just about 90% of your weight on your rear leg and snap off with the front.

Another piece of footwork we cannot diagram on the horizontal clock foot chart is the knee lift, the common kick or strike.

Another piece of footwork that cannot be diagramed on the horizontal clock is the "lifting foot," a counter to an incoming kick to the lead leg, be it snap, hook or stomp.

Another piece of footwork we cannot diagram on the horizontal clock foot chart is the stair climb. Stair shapes vary. The top of the stairs and thereafter really vary. Proper SWAT team and military assault team training facilities have a variety of stairs to climb and various floor plan problems once atop the stairs..

Some Points on Searching, Moving to Search and Footwork

Military manuals frequently suggest creating a "pad" for the stock of your long gun that is formed and set-up with a high elbow. Which...can be seen first creeping around corners. The solution - lower that elbow.

What else will the enemy see first? Coming around the bend?

When we plan on leaning over, our bodies have a propensity to place the toes of our feet just ahead of the lean, exposing the front of the foot. Don't you do this, and do look for this in searches.

Moving Off While Drawing and Shooting

Live fire shooters are often encumbered with moving drills and skills because of safety issues. If a line of shooters do a "step and shoot" drill, the instructor sometimes looks like a chorus line choreographer. You know the routine. Draw while stepping to the right and shoot. Draw while stepping to the left and shoot. With just a step, shooters are still maintaining a rather frontal, full-body, target offering to an enemy. Especially if they shoot with two hands – and not turning to run while drawing. If moving is key? If moving to cover is vital? Shouldn't we be busting off right or left like in a 100 yard dash?

These answers can be better examined with simulated ammo guns, interactively. But, once you do these "step-and-shoot" drills with an actual opponent shooting back at you, and the opponent is fairly close, you begin to realize that the idea is not as successful as conceptualized. The other guys still shoots you rather easily. This is not to say you shouldn't shoot and move, it's rather to say, don't expect miracles.

Miracles like…invisibility! Invisibility? There are very prominent instructors who still adhere to the cardboard tube theory of stress. They believe that when stressed out, adrenalized, their vision becomes blinding, tunnel vision. Essentially a cardboard tube.

They actually use the term "cardboard tube." And they teach that if you move aside, even just a step, you will become invisible to your opponent. They use the word "invisible." This cardboard tube idea has been disproven, and replaced by simple "target focus," and simple "attention focus" explanations. You are looking at, zeroing in on, what's important that second and your memory is not recording surrounding outside things and events for that second. Remember that researchers can only ask questions of your memory of incidents after the event, and base their "tubular" conclusions on your memory. You were not blind. You were just focused. The cardboard tube analogy is rather misleading nonsense and confuses reality and students.

Right-handed, American football quarterbacks are admired when they dash to the right and throw well "cross-body." Coaches say this is a special skill. Most times they can throw better as they dash to left and not throw cross-body. I think the same applies to shooting pistols. Will your opponent be right-handed (9 out of 10 people) and naturally choose to step or dash to their left and avoid the cross-body awkwardness? What direction will you "naturally" run in? Will you step or will you dash?

And by the way, a study of captured videos, body cams etc., show quite a number of shooters who DON'T move and just stand and shoot, and win, too! Get the sims guns out, get busy and experiment!

By the way, don't shoot yourself in the leg while drawing and moving!

Basic Unarmed and Armed Stand-up Footwork List

Use in your workouts, seminars or classes. Select sections as needed. Do sets of 5? Sets of 10? The instructor chooses the who, what, when, where, how and why.

* do unarmed or with weapons sheathed, holstered or mounted.

* do while drawing and holding a knife.
* do while drawing and holding a stick.
* do while drawing and holding a pistol.
* do while unshouldering/lifting and holding a rifle.

Set 1: Stand straight to bladed ready footwork.

Set 2: In and Out Footwork with a Center Foot Axis.

Set 3: V Footwork: Bladed with right foot forward.

Set 4: V Upside Down Letter V Footwork.

Set 5: The Thai Jump Leg Switch.

Set 6: Shuffle Footwork:
- 3 to axis to 9 and back.
- 5 to axis to 11 and back.
- 7 to axis to 1 and back.
- 9 to axis to 3 and back.
- 11 to axis to 5 and back.
- 1 to axis to 7 and back.
- 4 to axis to 10 and back.
- 6 to axis to 12 and back.
- 8 to axis to 2 and back.
- 10 to axis to 4 and back.
- 12 to axis to 6 and back.
- 2 to axis to 8 and back.

Set 7: Step and Slide Footwork.

Set 8: Quarter, Half and Full Circle Footwork.

Set 9: Pivot Footwork.

Set 10: Walking Lanes. Walking off the Clock.

Set 11: Run Off Footwork: Stand in the axis and then:
- dash off at 12.
- dash off at 1.
- dash off at 2.
- dash off at 3.
- dash off at 4.
- dash off at 5.
- dash off at 6.
- dash off at 7.
- dash off at 8.
- dash off at 9.
- dash off at 10.
- dash off at 11.

Set 12 Combinations, Do All:

Segment 3: Knee Maneuvers

Take a Knee...or Knees!

How often do you end up on your knees? There are unplanned times, like with sudden large explosions or the reflexive need to duck down. There are times when you get behind cover or concealment. The knee down comes in three studies.

1: right knee down
2: left knee down
3: both knees down

1: knee versus standing enemy
2: knee versus knee-high enemy.
3: topside in ground fight.

Sounds simple but you may have to move around on your knees after dropping. Not crawling yet. Crawling is a separate study coming later. The following exercises will cover "knee maneuvering."

Knee versus standing.

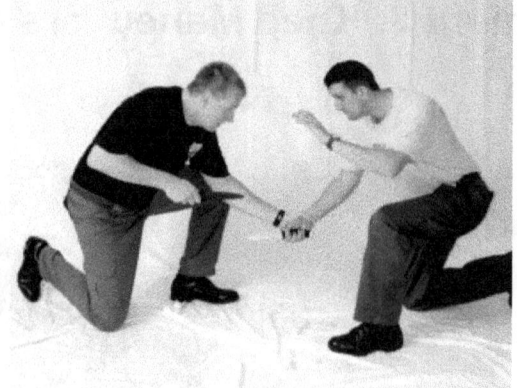
Knee versus knee.

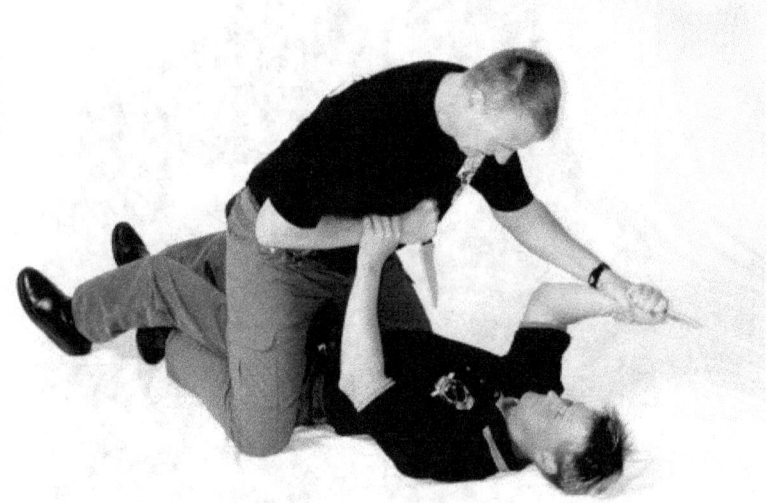
On your two knees topside in a ground fight.

Maneuvering on your knees can happen at any time.

A Knee Maneuver Stretch Exercise
Before we start with the patterns, here is a bent knee exercise for stretching.

1: Starting, neutral point.

2: Right foot steps out, then back.

3: Left foot steps out, then back.

4: Right leg shoots out to 1:30 o'clock, and then back to neutral.

5: Left leg shoots out to 10:30 o'clock and then back to neutral.

6: Right leg shoots out to 3 o'clock, and back.

7: Left leg shoots out to 9 o'clock, and back.

7: Right leg shoots to 4:30 o'clock, and back.   8: Left leg shoots to 7:30 o'clock and back.

Limited in comparison to other maneuvers, "traveling" on the knees, stepping off to one knee, one foot, is an important, possible occupance. As we have discussed earlier, the ground, floor, terrain is always an issue. The world is not matted and this may or may not hurt a bit.

Knees Number 1: Squat Down. Be aware that sometimes you can just squat straight down from standing. Once down, shift your leg and body a bit to get a comfortable knee position. No official footwork needed with this. It is a standing-to-knee high, drop-down movement.

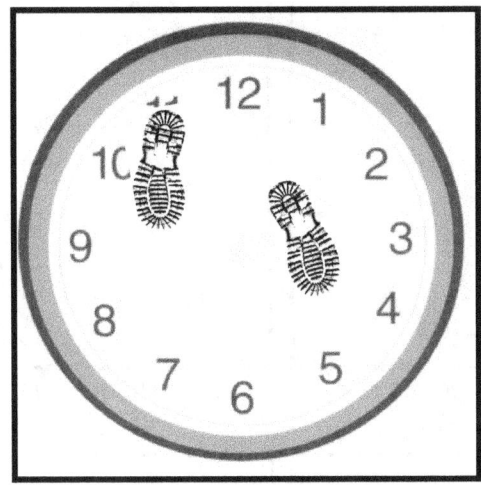

Standing.

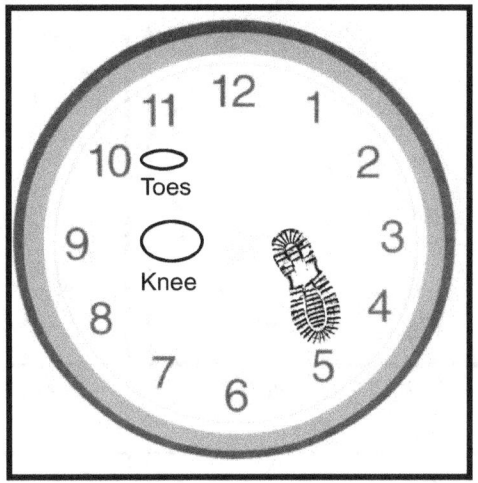

Drop torso straight down to take a knee. Either leg.

Many citizens, police and military can just drop straight down into a kneel, as well as stand straight up. Some can't. You should experiment with this.

Knees Number 2: Knee-Off. Both knees are down. To maneuver and remain balanced, knee "step" directions are limited. The right knee or the left knee "steps" off into a direction. The other knee may or may not follow. Depends on the situation. The moving knee will only travel within a cone because of balance.

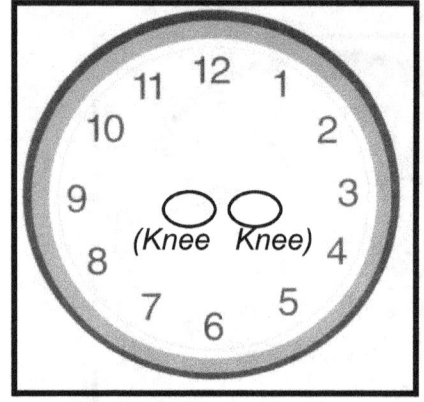
1: A neutral position.

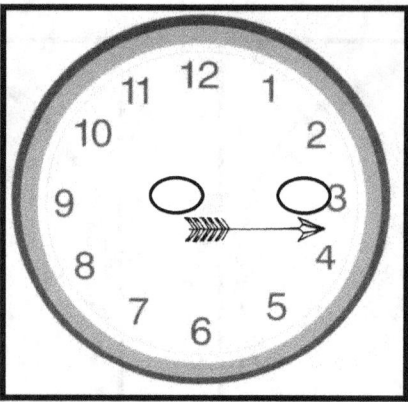
2: Right knee moves out.

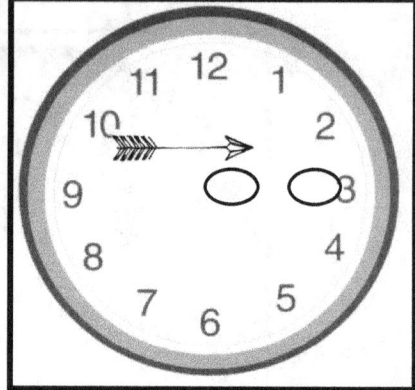
3: Left knee may follow.

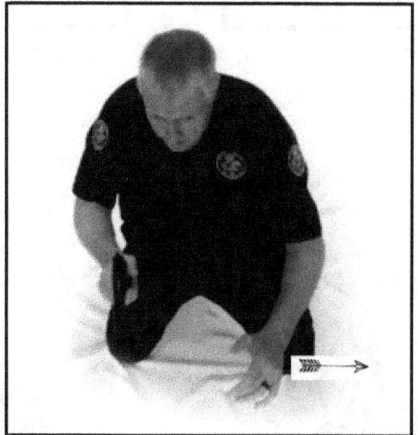

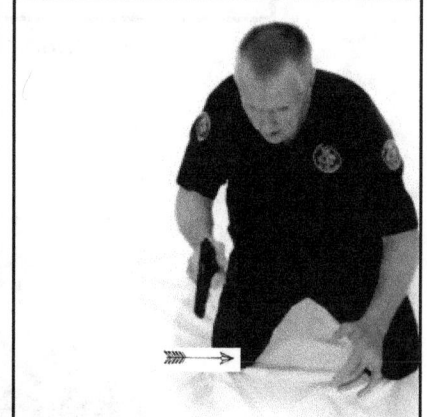

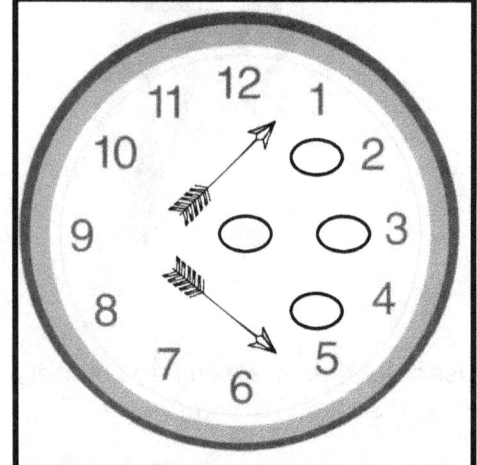
The right knee will probably only travel from maybe 1 to 5 on the clock. Balance.

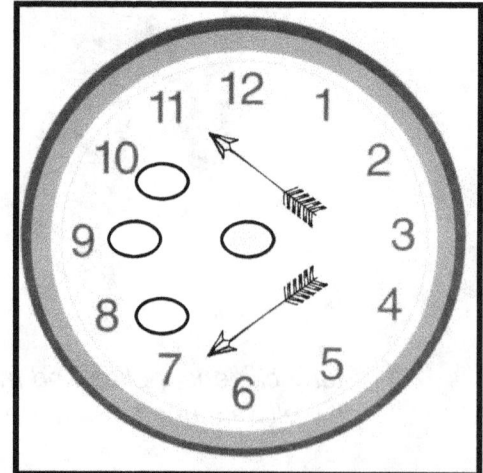
The left knee will probably only travel from maybe 7 to 11 on the clock. Balance.

Knees Number 3: Knee-Step Off to a Knee Up. Like the knee/leg stretching routine shown a few pages prior, with a foot step-off you go from two knees down to taking a right or left step and having one knee up. See if you can do two "knee steps." Or more? If not, further low maneuvers turn into crawling.

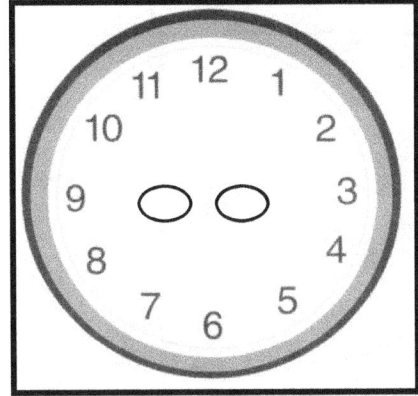
1: You are in a neutral, two knee down position.

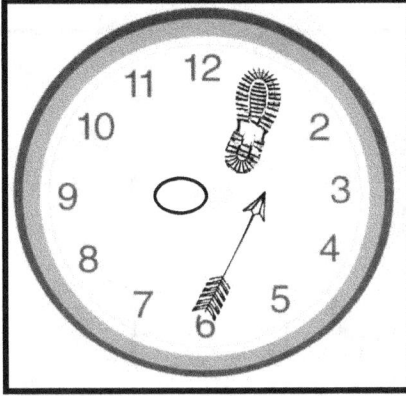
2: The right foot steps off.

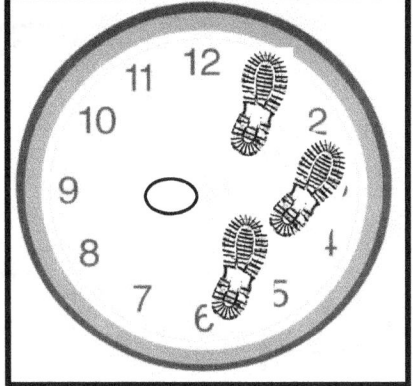
3: The right foot can only step off with a cone of balance.

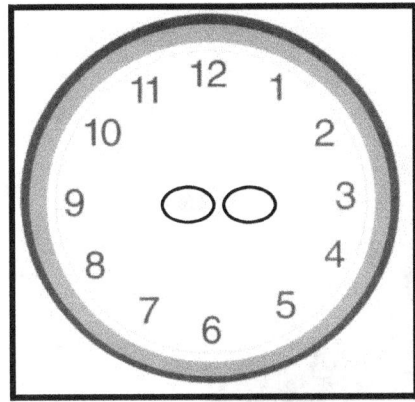
1: You are in a neutral, two knee down position.

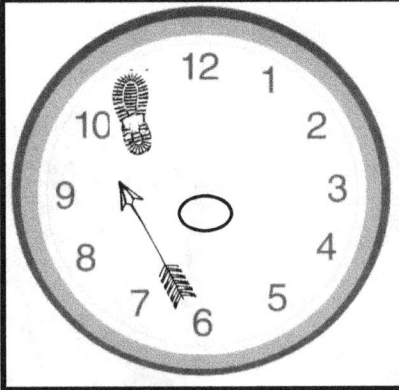
2: The left foot steps off.

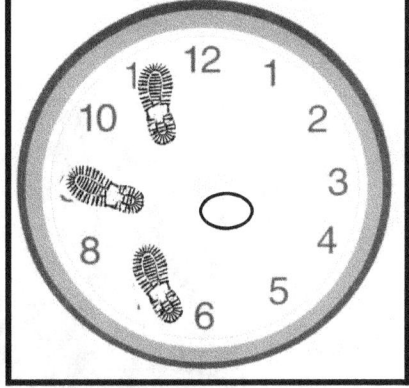
3: The left foot can only step off with a cone of balance.

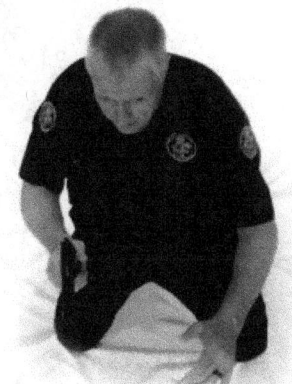

Knee stepping forward or back, a healthy distance between the knee and the foot creates stability. See next.

Knees Number 4: The Side-to-Side, Torso Glide. With one knee up and one knee down, you have three points touching the ground. But if your foot is close to your knee, you have less stability and could even be knocked over to the side. If your foot is away from your knee it creates stability and allows your torso to glide side-to-side a bit, say to peer around a corner or a door, a building or a car.

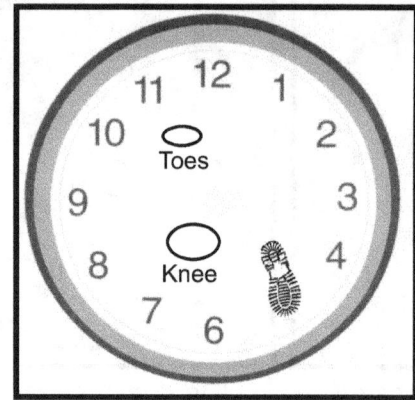

The knee is close to the foot.

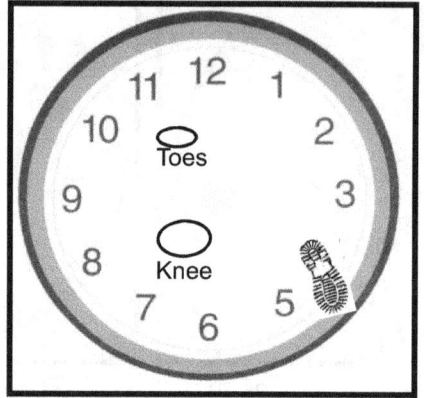

The knee is not too close the foot.

A wider berth allows you to shift right and left a bit. And of course, does make for a bigger target if not behind cover.

Knee Notes 1: Flip a Table. Go Knee High

I was eating at a restaurant once reading some gun magazines. One article's theme was shooting from a table and chair situation. Not unlike the very table and chair where I sat at the moment. The purpose was to teach folks how to draw and shoot under stress while first seated at a table.

The magazine photo series started out with a classic outdoor "square" shooting range. Sunny day. About eight guys were seated in chairs on the firing line. Each one had a table in front of him. Both the tables and the chairs appeared to be cheap, like folding tables and folding chairs.

Obviously upon some signal, the men stood up from their seated positions and formed two-handed grips and shot the paper targets. I could see in the photos that one of the chairs had tipped over. All the tables were still standing, and all the men loomed over the tables like giant targets themselves, full-frontal to the imagined "threat" they were in a gunfight with. Did the range owner not want to damage his tables? Why not consider flipping the table over and drop behind it for cover while drawing and shooting as a part of the training day on this subject?

What might a flipped-over table do for you? How about some possible cover? Or at the very least, some confusing concealment? What about stopping and/or deflecting various types of bullets. Will every caliber bullet go through every table on the planet? What is the bad guy's gun caliber? Will your table offer you some cover? Even some concealment? Would some upturned tables stop some rounds? Slow them down? Would tilted tables, tilted at least from the bad guy's perspective, cause his bullets to deflect/bounce rather than pierce straight into you? Think of all the kinds of tables. Or would we, should we just completely forget the flipping idea and stand up there all tall and looming over an upright table like a giant silhouette target?

The table I sat at was a heavy metal table. I say, "Flip up and drop down," while drawing and shooting from the seated "table-and-chair" scenario. In the "who, what, where, when, how, and why of life? Next time you are in a restaurant? Take a quick look at your table. Get ready to flip and drop.

Knee Notes 2: The Self Stab, Self Shoot

There are two kinds of knife grips, the saber grip and the reverse, or ice pick grip. Both have many advantages. And disadvantages. One of the disadvantages of the reverse grip is that the knife tip is easily aimed right back you. Accidents can cause self stabs. For example, with the raised knee on the knife hand side, take care not to stab your leg, or stab anywhere when ground maneuvering.

The ice pick grip is a self stab threat.

Keep alert to your ice pick, grip tip on all ground maneuvers.

Don't shoot your thigh in a knee high draw!

Knee Notes 3: The Ankle Trip

Police and military team members should know that when a member of the team has one knee down, there is a chance for "hook" another's foot to trip on a high ankle.

This is a tripping hazard for team members.

This is less of a tripping hazard for team members.

Basic Unarmed and Armed Knee Maneuver List

Use in your workouts, seminars or classes. Select sections as needed. Do sets of 5? Sets of 10? The instructor chooses the who, what, when, where, how and why.

Set 1: The Knee Stretch Exercise
- do unarmed or with weapons sheathed, holstered or mounted.
- do while drawing a knife, a stick, a pistol, a rifle.
- do with weapons presented.

Set 2: Standing to Drop Down Squat
- do unarmed or with weapons sheathed, holstered or mounted.
- do while drawing a knife, a stick, a pistol, a rifle.
- do with weapons presented.

Set 3: Knee Routine Number 1: Knee-Off
- do unarmed or with weapons sheathed, holstered or mounted.
- do while drawing a knife, a stick, a pistol, a rifle.
- do with weapons presented.

Set 4: Knee Routine Number 2: Knee-Step Off to a Knee Up
- do unarmed or with weapons sheathed, holstered or mounted.
- do while drawing a knife, a stick, a pistol, a rifle.
- do with weapons presented.

Set 5: Knee Routine Number 3: The Side-to-Side Torso Glide
- do unarmed or with weapons sheathed, holstered or mounted.
- do while drawing a knife, a stick, a pistol, a rifle.
- do with weapons presented.

Segment 4: Standing to Ground Maneuvers

Going prone. You drop down in the four basic, generic "clock" directions, to the front, to the back, to the right. To the left. And to points in between. Whether suddenly or gradually.

You drop:
    1: Somewhat to the front, on your chest.
    2: Somewhat to the back, on your back.
    3: Somewhat to the right side, on your side.
    4: Somewhat to the left, on your side.

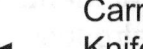

Unarmed.
Carrying but not drawn.
Knife in hand.
Pistol in hand.
Long gun in hands.

Sometimes you drop in a progression. You roll into the drop. Sometimes, some athletic people can leap down in "one step." I have seen a person, respond to a surprise explosion so fast, he dropped face down so fast, he cut his chin open on the ground.

Keep in mind that when you drop down you may be unarmed, or armed with a baton, a knife, pistol or a long gun. This will inhibit you from using two hands to help your descent. Armament may cause the support limb into a one-hand, one forearm response. Many release a single, non-trigger hand from the grip of a long gun and descend. Here are some descending progressions:

Progression 1: Standing/walking/running. Squat to two knees to chest down.

Progression 2: Standing/walking/running. Squat to one knee to chest down.

Progression 3: Standing/walking/running leap to two hands/forearms to chest down.

Progression 4: Standing/walking/running leap to one hand/forearm to chest down.

Progression 5: Standing/walking/running leap to two hands/forearms to one or two knees to chest down.

Progression 6: Standing/walking/running leap to one hand/forearm to one or more knees to chest down.

Progression 7: Dropping backward. Squat. Some people fold one leg behind them and use that as a stepping stone on their way down to their back.

Progression 8: Dropping backward. Squat and "sit" as best they can and roll back.

Progression 9: Dropping to the side, whether left or right. Some people can arch their body, curve it to roll down to their chosen side.

Progression 10: Dropping to the side, whether left or right. Some people can arch their body, curve it to roll down to their chosen side and put out a support limb to help the descent.

Hitting the Deck! Accidentally or on Purpose? Fall or Drop?
According to paratroopers, stunt professionals, physical therapists and martial arts instructors, there is a "better" way to hit the deck. Experts said you save your elbows and knees and try to take the hit on your "meat." Kate Murphy of the *New York Times* collected some advice from experts about accidental falls, advice which also relates to on-purpose drop-downs also.

"Aim for the meat, not bone," said Kevin Inouye, a stunt man and assistant professor of acting, movement and stage combat at the University of Wyoming. "Your instinct will be to reach out with hands or try to catch yourself with your knee or foot, but they are hard and not forgiving when you go down."

The key is to not fight the fall, but just to roll with it, as paratroopers do. "The idea is to orient your body to the ground so when you hit, there's a multistep process of hitting and shifting your body weight to break up that impact," said Sgt. First Class Chuck Davidson, master trainer at the Army's Advanced Airborne School at Ft. Bragg, N.C. Of course, you will be better able to loosen up, pivot to your side, tuck and roll if you are in good physical condition.

"If you have a room full of soccer players and computer desk workers and go around knocking people over, you can bet the soccer players are going to be less likely to get hurt because of their superior strength, agility and coordination," said Erik Moen, a physical therapist in Kenmore, Washington.

Jessica Schwartz, a physical therapist in New York City who trains athletes reports "It's almost inevitable you are going to fall, so you really should know what to do. 'The number one thing to remember is to protect your head. So tuck your chin. In planned descents, especially with firearms, your eyes should be on the enemy or the enemy's area.

On Mats...

All of experts will warn that not reflexively extending your hand out when falling or dropping is a challenge. Some say this is an injury waiting to happen. Others suggest using your hand, forearm and arm to slow or control the fall. I think is it situational. If you are planning a descent then your limb can be put to good use. Planned or not, you can hurt your hand and arm.

Somewhere between the accidental fall and the premeditated descent is the argument for or against slapping the mat. Slapping is supposed to reduce the impact. It is thoughtless and popular with many martial artists on mats.

On mats? Decades ago, I took my 3rd Dan AlkiJitsu test. Just me, another candidate named Dennis and the head guy, Dr. Carl Fareneali (a real college professor and actual doctor). While a martial camp ensued outside we entered a building for the test. It had a cement floor. We pulled mats out. The object of the test was to execute a series of strikes and kicks, takedowns and throws. We had to demonstrate the throw or takedown and also be tossed and dropped by Dr Carl. Since it was an upper dan black belt test, their was much time spent airborne.

At one point, Dr Carl threw Dennis through the air. Carl had a foot on one mat and a foot on another mat. And the throw began, his feet split the mats apart, creating a "V," an opening of cement on the floor between the mats. Dennis landed. As required he slapped. His right hand and forearm slapped the mat. No problem. His left hand and forearm slapped the cold cement floor. Problem.

He screamed out in pain. We all investigated. At first he thought he had broken his arm in some manner. But he could twist his wrist. He could move all his fingers. He could bend his arm at the elbow. But it all hurt. While this was not the 1950s and 60s when all "Kuraty" was especially vicious, laced with blood and vomit, it was the 1990s and he still had to continue. He, I and others, when taking tests and getting injured in them, were used to the expression, "you know, this test isn't over yet." We continued.

For me this was a little scientific test about slapping, right before my eyes. One limb slapped a mat. The other limb slapped cement. The primitive thought came to my brain, "slapping not good." Not good for the "street" as they say. Floors, roads, sidewalks.

I looked for alternatives and started re-liking the rolling of Silat and even ninjitsu. Other system roll rather than slap too. And I worked on the Parachute Land Fall, a sort of vertical roll all unto its own. I never slapped again. Never will. My hand may instinctively still shoot out with a surprise fall. But not on purpose.

Another point I will make about slapping. After many repetitions of falling and slapping, this creates the habit of doing so. Then, when a enemy throws you down, you land, slap the pavement, possibly debilitate your limb and leave you, by your own accord laying prostrate at the feet of your enemy. If you rolled away instead? You would be escaping his next moves, one of which could be a face or neck stomp.

I will never slap or teach slapping for the non-mat world. Round off all your edges and roll. You will still find some military manuals showing slapping. They have been unduly, mindlessly influenced by matted, martial arts. And, some indoctrinated people will never agree and never give up these mandatory slaps. Of course, if you have a gun, a knife or a stick out, you can't slap to martial arts standards anyway.

Parachute Instruction. The last 21 feet can be tricky and this is the recommendation.

Shooting on the Ground

Dropping to the prone, chest-down, or supine back positions, or right side or left side down positions for shooting a pistol or a long gun requires a level of fitness, plus dealing with an eye disorientation as well because of a bounce and a different cant of the weapon (and sights) for shooting. Lots of jumping up and down and dry-firing is required for this training, which many people find difficult, or ignore.

Getting flat for civilians, enforcement, police and military could be for cover and, or concealment. People with a pistol do not usually go prone to slowly squeeze off a precision, "sniper-style" shot. Rather they are responding to a hot situation inside a particular landscape. Long gun? You might drop down for a precision, supported rifle shot.

In almost all environments, suburban, rural or urban many "things" offer more bullet resistance closest to the ground. Usually trees are thickest near the roots. Most tall man-made objects have their architectural strength closest to the ground.

There are numerous true stories about successful shooters dropping down in crime fighting and war. Too numerous to mention here, and here is not the place for such a list. Research them to learn the various circumstances.

On the descent, try to keep your eyes and the weapon up and at the enemy. Use the non-trigger support hand as a tool to lower yourself.

Get Down 1: Standing (to a Knee or to Two Knees) to Prone with Two Hands. Legs Out. If you are empty-handed or your weapons are holstered, sheathed, pocketed, mounted with a shoulder sling, etc. this is a most common way to get prone. It can resemble the classic burpee exercise. To get back up you reverse the steps.

Start standing.   Squat as far as you can. You might even get on your knees. Use hands.   Put hands down solid and shoot feet out.

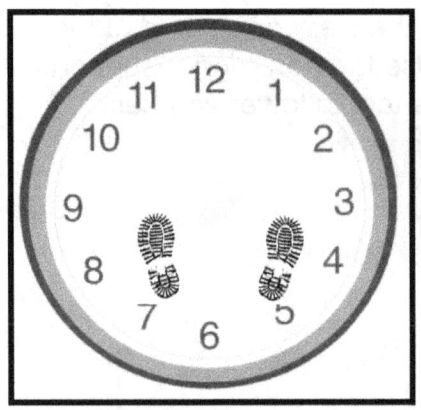
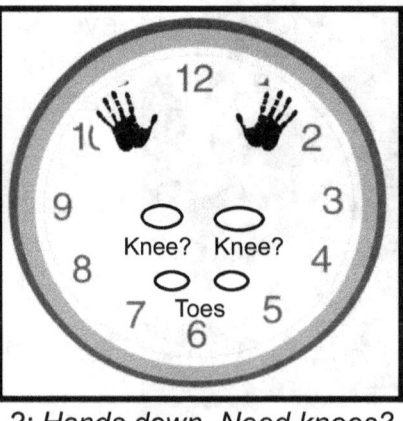

1: Standing. You see the need to drop.   2: Hands down. Need knees? If not...   3: ..."shoot" feet out.

Page 78

Get Down 2: Standing to a Knee or to Two Knees to Prone with One Hand Help

1: You see the need to drop!

2: In this sample, squat as far as you can, a leg shoots out, a hand goes down...

3: You descend.

Prone shooters often opt to roll half-right or half-left for comfort.

Keep the firearm aimed at the enemy, ready to shoot, on the descent.

Get Down 3: Standing to a Rolling Side Drop Probably the only reason you would drop to the right or left side in this manner without turning or twisting your torso would be because you are holding a firearm. Athletic people can arch their bodies and sort of roll down. Then, rock for momentum and roll back up again enough to get the down-side leg bent and under them to completely get up. If you are not shooting, then remove your finger from the trigger. Many people cannot and should not do this.

1: You see the need to drop to the side!

2: You can squat as far as you can, touch ground maybe, maybe not, and roll the rest of the way.

3: Right side or left side. Same process.

A sample of an athletic roller.

Get Down 4: Dropping Supine The common consensus is that you again squat. You try to sit on your butt as much as possible, as quickly as possible and then roll back. Don't let your head hit the floor. If you are armed, you will try to keep your eyes and gun aimed at the enemy. If you are not shooting, then remove your finger off the trigger.

1: You see the need to drop back.

2: If possible, you squat, sit back and roll back. Don't let your head hit the floor. Don't shoot your legs.

3: If you spread your legs, they may collect incoming rounds and bounced rounds.

4: Crossed ankles prevent this bigger target problem. Don't shoot your feet!

Gun people will argue about how you should position your legs when you land with your firearm in a gunfight. Legs spread? Or ankles hooked. It is easier to roll over with your ankles hooked. You also have a smaller target profile. Choose for yourself.

Basic Get Grounded, Unarmed and Armed List

Get Down 1: Standing to a Knee, or to Two Knees, to Prone with Two Hands

- do unarmed or with weapons sheathed, holstered or mounted

Get Down 2: Standing to a Knee or to Two Knees to Prone with One Hand

- do unarmed or with weapons sheathed, holstered or mounted.
- do while drawing a knife, a stick, a pistol, a rifle.
- do with weapons presented.

Get Down 3: Standing to a Rolling Side Drop

- do unarmed or with weapons sheathed, holstered or mounted.
- do while drawing a knife, a stick, a pistol, a rifle.
- do with weapons presented.

Get Down 4: Dropping Supine

- do unarmed or with weapons sheathed, holstered or mounted.
- do while drawing a knife, a stick, a pistol, a rifle.
- do with weapons presented.

Can you really do this like a cartoon hero? Some can. Most can't. Some must.

Segment 5: Crawling Maneuvers

Ground fighting and maneuvering includes crawling. If you are on your knees, and/or feet, and your hand, hands and, or forearms go down to help move, you are officially crawling. Crawling is a category and study unto itself, near and dear to the firefighting, military, exercise programs and obstacle courses. Some crawling can be a challenge experts say, when it requires contralateral movement, that is moving your right arm and left leg forward at the same time.

Pennsylvania Firefighter and Fireman Trainer Tim Llewellyn reports, "There are solid reasons why helmeted firefighters are encouraged to crawl in a structure-fire environment:

    1: It's cooler down low.
    2: Visibility is often better nearer to the floor."
    3: And needless to add, fewer burned chemicals in the smoke above.

Military and police learn to crawl to remain hidden from the enemy and to counter common height gunfire.

Exercise programs like to promote crawling for fitness and strength benefits.

Obstacle courses include crawling through pipes, half-hoops, mud, water and just about anything they can muster for a course.

An Italian school runs a fire drill, teaching kids to escape the dangers of smoke and all that smoke entails.

Page 83

Crawl Number 1: The Face Down Crawl There are three kinds of face down crawls, a low crawl, a high crawl and a higher *bear crawl* (explained later). You pick the one you need for the situation and the moment. Low is when your chest does touch the ground. High is when your chest does not rub/touch on the ground, but your knees do. The bear crawl is higher still. We have been crawling since infancy and do thus naturally. But for how long?

The newer challenge is crawling side-to side and backward. You will figure this out the first seconds that you try. The exercise here is to crawl off in the directions of the clock, as called off by a trainer.

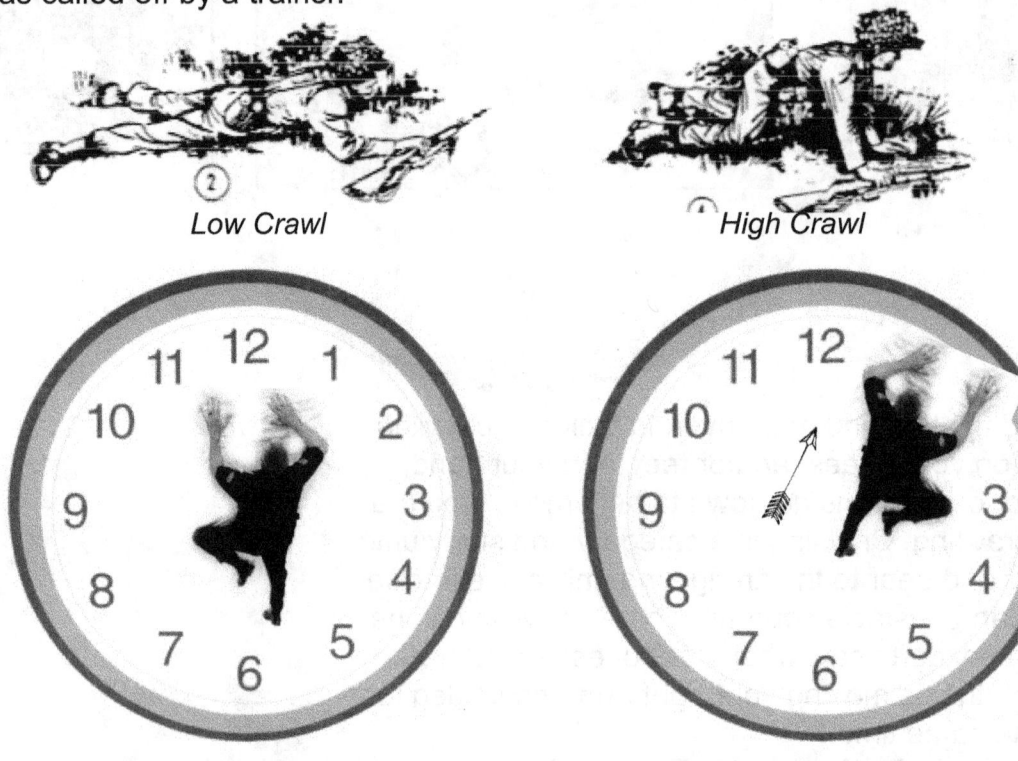

Low Crawl   High Crawl

1: The trainee gets into crawl position, assigned high or low crawl.

2: The trainer calls out a clock direction.

The highest crawl is called the *bear crawl* (to the left) and can be popular in exercise programs and group work-outs for civilians and the military. Your knees are off the ground, and only your hands and feet touch the ground.

Crawl Number 2: The High Crawl The trainer calls out a clock number and the trainee crawls off into the direction.

The trainer calls out a clock number.   The trainee responds.

Crawl Number 3: The Circle Crawl The trainee crawls in a circle clockwise, then counter-clockwise.

Sometimes you have to crawl sideways in parts of a circle, or perhaps even a whole of a circle. Clockwise or counter-clockwise.

Crawling Notes

As defined by the militaries, firemen, and reputable exercise programs of the world, here are some worldwide training manual diagrams and photos demonstrating the subject of face-down crawling for inspiration.

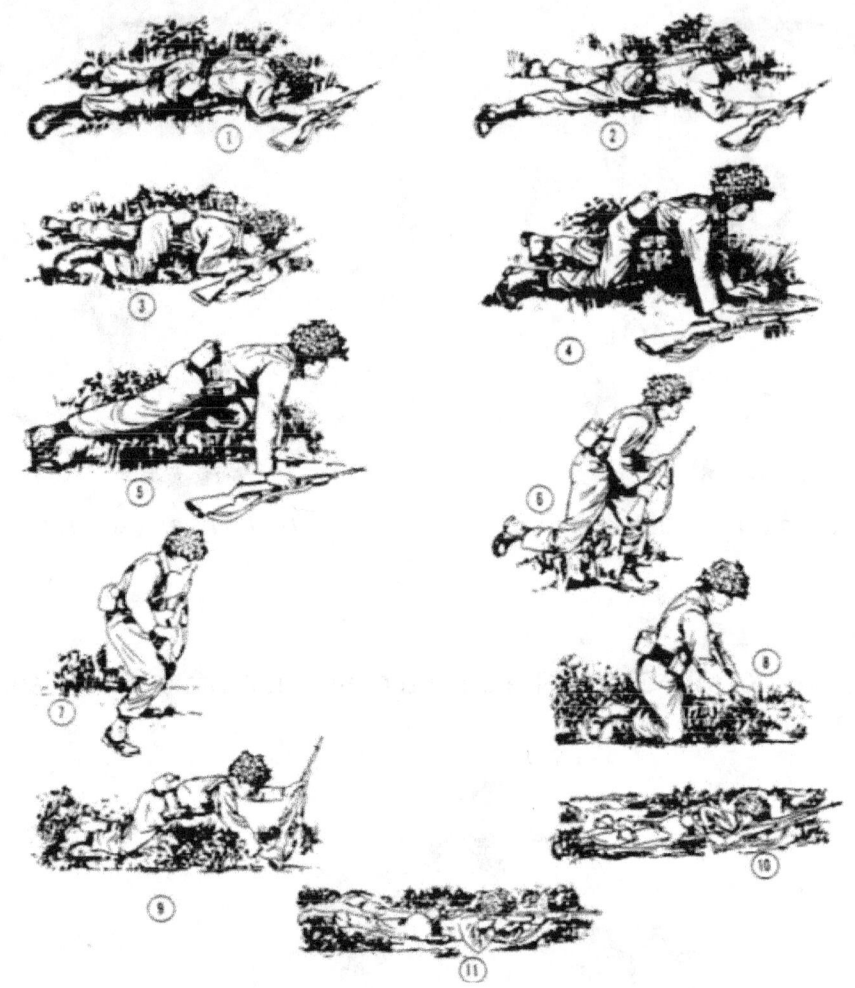

The one hand sling carry, weapon on arm.

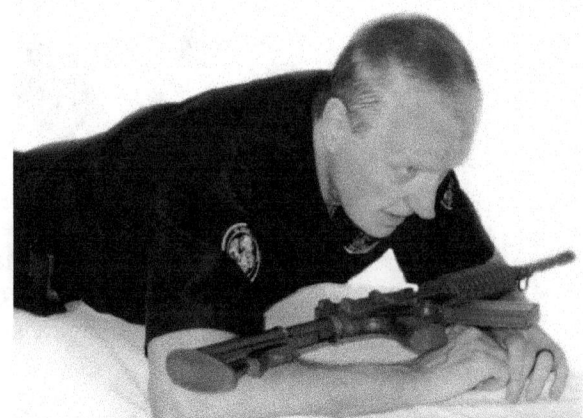

The cradle carry, weapon on arms.

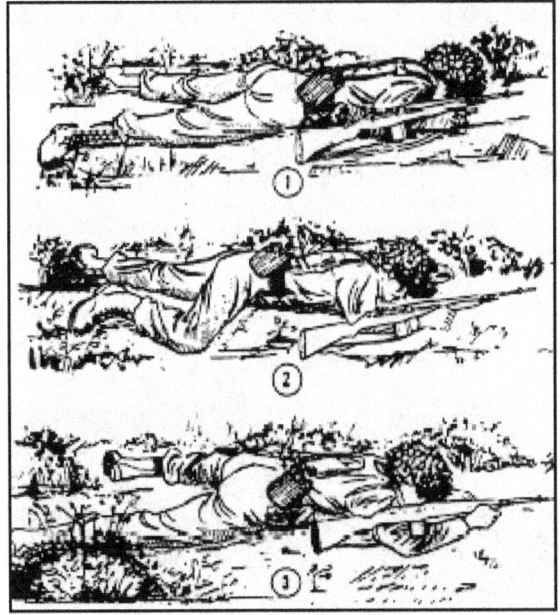

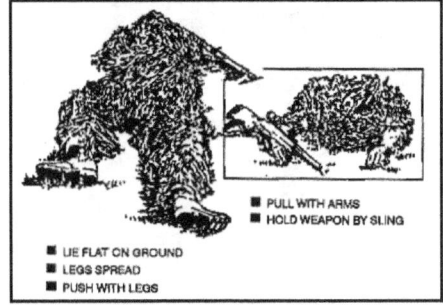

Crawl Number 4: The Knee Drag Crawl There are two kinds of knee drags. Shoulders mostly down, or one shoulder mostly up. Your position your hand or hands, and foot, dragging the rest of the body. You head can be up and looking around. This is an "old-school" part used in approaching a sentry for a sentry kill; however this can be can be very noisy depending on the terrain. This might also come into play with a wounded leg.

Shoulders mostly down.   *One shoulder mostly up.*

The aforementioned firefighter Tim Llewellyn reports, "Certainly, crawling on hands and knees offers a few positive points: we've been doing it since we were infants, so it requires no explanation; and, transition from standing or kneeling to crawling is relatively simple and efficient."

He continues, "The improvement I've learned is a head-up, leg-out, or tripod position; the variation is slight, but our firefighters have found the benefits to be numerous. Instead of crawling on both hands and both knees, in this technique, one knee is down on the ground, the other knee is up with the foot outstretched, and one arm outstretched, hand down on the ground. Advancing entails reaching out with the down hand, stepping forward with the up knee, and dragging the down knee behind as the body moves forward to maintain the tripod position. It frees one hand to hold a tool."

In the clock maneuvering exercise, a trainee gets down in a knee drag position. A trainee calls out a clock number direction and trainee drags a leg in that direction.

Crawl Number 5: The Crab Walk Crawl Another favorite exercise, and I have come across this movement a few times in fights, where a few crab steps were important for positioning. Begin by sitting on the floor with your feet hip-distance apart in front of you and your arms behind your back with fingers facing hips. Lift hips off the floor and tighten your abs. Start "walking" forward by moving your left hand followed by your right foot; and then your right hand followed by your left foot. Walk four or more steps as space allows, then walk back. Continue back and forth for desired amount of time.

In the clock maneuvering exercise, a trainee gets down in a crab crawl position. A trainer calls out a clock number direction and the trainee crab walks in that direction.

Trainee gets ready.

Trainer calls out a clock number.

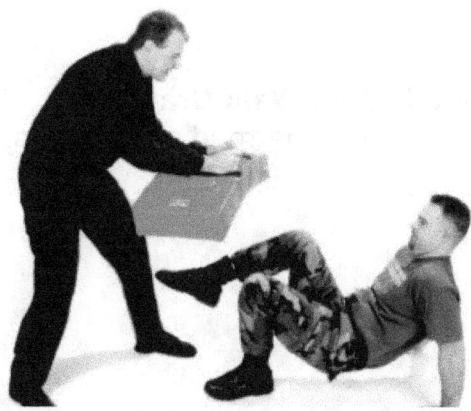

One very useful application is getting into quick positions for ground kicks, moving for the best delivery spot. Also the lift helps generate kicking power.

Basic Crawling, Unarmed and Armed List

Set 1: Face Down Crawl: Low
- do unarmed or with weapons sheathed, holstered or mounted.
- do while drawing a knife, a stick, a pistol, a rifle.
- do with weapons presented.

Set 2: Face Down Crawl: High
- do unarmed or with weapons sheathed, holstered or mounted.
- do while drawing a knife, a stick, a pistol, a rifle.
- do with weapons presented.

Set 3: Face Down Crawl: Bear Crawl
- do unarmed or with weapons sheathed, holstered or mounted.

Set 4: Circle Crawl
- do unarmed or with weapons sheathed, holstered or mounted.
- do while drawing a knife, a stick, a pistol, a rifle.
- do with weapons presented.

Set 5: Knee Drag
- do unarmed or with weapons sheathed, holstered or mounted.
- do while drawing a knife, a stick, a pistol, a rifle.
- do with weapons presented.

Set 6: Crab Walk Crawl
- do unarmed or with weapons sheathed, holstered or mounted.

Segment 6: Universal Grounded Maneuvers

These are the "on your back and sides," grounded universal moves I think everyone should know from my 40 years-plus of martial study as well as numerous non-sport ground fights as a police officer, plus my investigations of crimes.

By universal, I mean every good system has them as core essential maneuvers and positioning. If they don't? They should. What you chose to do after the positioning is left to your lifelong study to find your favorite follow-ups.

Grounded Number 1: Ready on the Ground We will start with something super simple, and very important. Through the years as I teach all kinds of people, I always include ground fighting. When a novice ground fighter gets on the ground, they tend to lay flat, toes up. When a person with some martial ground experience gets on the ground, they tend to put the soles of their shoes on the ground. They know that they are in a better action position with their feet braced in this manner. I can do a quick gauge on the experience of the class by looking their feet over.

A great place to start is this knees up ready position. Their arms should also be up and at-the-ready when doing these ground exercises too.

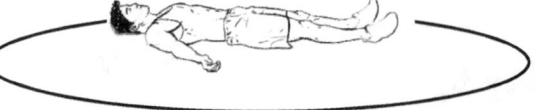
Some exercise trainers call this the "corpse" for good reason.

Knees up is a better ready position to begin these exercise and action.

Grounded Number 2: The Shrimp, or Hip Escape This bottom side movement is called the "shrimp" because of the shape your body makes. In a fight you pick the best side to push off on, that being the side you can shove your opponent off of you and into.

Shove him off to your right? Put your left foot up near your butt. With great explosive, bench-press like power, shove your hands to his left, shove your butt out to the left.

Shove him to your left? Put your right foot up near your butt. With great explosive, bench-press like power, shove your hands to his right, shove your butt out to the right. Your body should end up in a really curved position, like a shrimp.

Wrestlers will want to roll up on their overturned opponent. "Street survivors" will not roll up, opting to get up or perhaps kick his face or body.

He's atop of you. Pick a side.   *Shove. Really slip that butt out into a shrimp.*

Shrimp out to the 10 and 2 o'clock positions,

Sometimes your butt goes up to the 2 o'clock or 10 o'clock position. It's situational. Sometimes, you use the shrimp concept, but "slowly." You slowly have to work or inch out in the 3 or 9, 2 or 10 directions.

If you wish to learn more subtle nuances in "the shrimp," you can study arts like Brazilian Jujitsu.

Grounded Number 3: The Shoulder Walk He's hit you, or shoved you. You've hit the deck. You're on your back. He's still afoot and moving in. To gain space (and time) you shoulder walk backward so you can kick him, or draw your weapons. It's like a crawl but it's face up and flat. This move could go into the prior *Crawl Segment*, but I decided to put it here. You use your shoulders and your feet to move back.

A shoulder walk may allow you enough time and space to get to your weapon, or start kicking, or perhaps even get back up.

Grounded Number 4: The Bucking Bridge The bridge in groundfighting. By definition, "The bridge is a grappling move performed from a supine position. All fighters need a strong neck, especially for certain bridges. There are two definitions/missions/uses for the "martial" bridge.

In many wrestling arts, you lose the match when your shoulders and, or back are pinned to the floor. A pin is typically defined as "a victory condition in various forms of wrestling that is met by holding an opponent's shoulders or scapulae (shoulder blades) on the wrestling mat for a prescribed period of time. One way to thwart this is to lift the shoulders and back off the mat - via a bridge with the neck and head. Arching your neck, creating a bridge can interrupt that. This is dangerous to your neck.

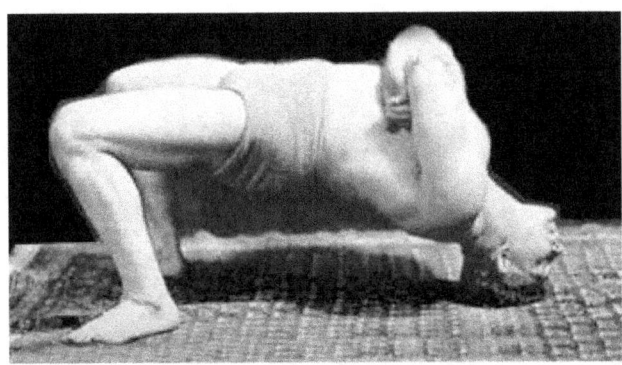 

The wresters, MMA and "reality" defenders often use the neck bridge, sometimes combined with a twisting body motion, to dislodge or flip off an opponent who has established a position on top.

But off the wrestler's mat where there is no automatic "back pin" loss, and a fighter needs to dislodge an attacker atop him or her without the extension of the neck in such a precarious and unprepared condition. The bridge concept can be a sudden, bucking motion from just the shoulders, not involving the neck. This lift may buck the attacker forward, and you continue this with a shove and, or the aforementioned shrimp escape and possibly the use of ground n' pound and grappling methods.

Bridging can be rolled onto just one shoulder, which, depending on the shoulder used, partial turns you to the right or left and throws him to over to one side.

Champion MMA fighter Bas Rutten describes being atop a skilled fighter (as in the photo to the left) can be like riding a surfboard in a storm at times.

A bucking bridge usually requires follow-up work.

A bucking bridge.

With perhaps a shrimp follow-up.

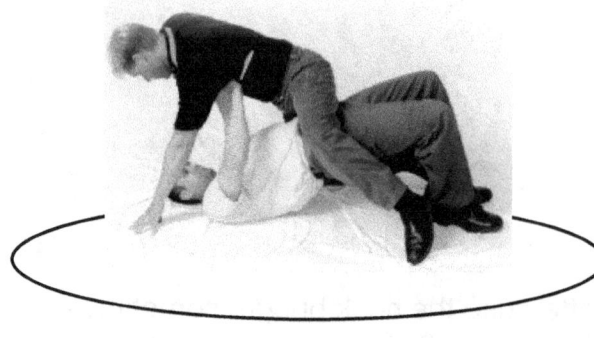

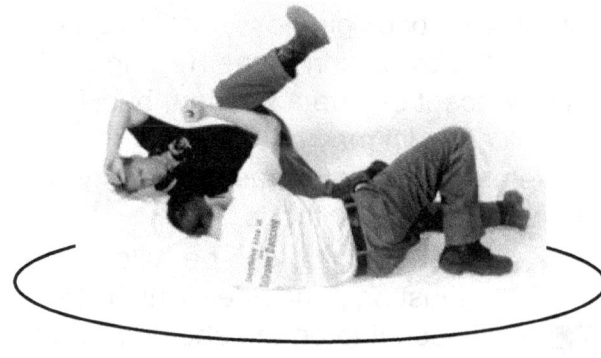

You can sometimes push with a baton, long gun, or a knife and a support hand.

Grounded Number 5: The Fishtails (Head and Torso) A fishtail flips/flops side-to-side. Sometimes you need to, too. Dodging ground n' pound punches is important when on your back. You can block if you can, but part of a block is also getting your head out of the way, or the skill of "fish-tailing" your head side-to-side and fishtailing your torso side-to-side if its free to do so. (The "art" and science of blocking appears in other books and films.) The exercise with this is to dodge controlled strikes executed by a topside trainer.

1: You see incoming. *2: Your head fishtails to the best side.*

1: You see incoming. *2: Your torso fishtails to the best side.*

Grounded Number 6: The Sit-Up There's the standard sit up and twisting sit up. Exercise gurus like to argue about the value of the sit up. Whether it destroys your back in repetitions or not, a ground fighter will have to on occasion, do something that looks like a sit up and/or a twisting sit up. And do it with explosive power. How you develop this is up to you.

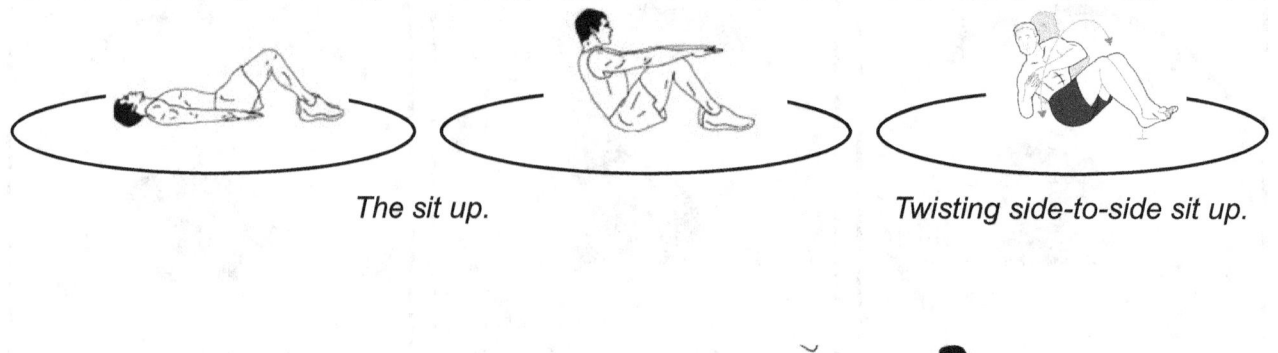

The sit up.   *Twisting side-to-side sit up.*

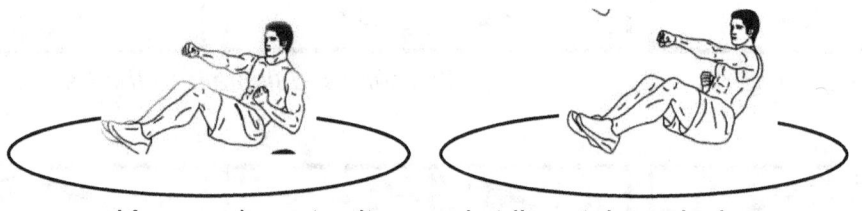

You may have to sit up and strike, stab or slash.

We are all familiar with the classic sit-up. You may have to sit up and strike, unarmed or with a weapon. You may have to sit up and grab a head, neck, torso, legs, or arms with your two limbs or just one limb. Then smother it, or push it or pull it.

You may have to sit up, twist and reach, grab and even pull back down, right or left sides.

Grounded Number 7: Guard Rotation This is a good exercise. Hook your ankles. Make a circle with your ankles. Each corner is a strategic move that could manipulate a topside opponent between your legs into a safer, strategic position for your plans.

1: Pushing back and away, pushes your topside opponent back.

2: Pulling in, pulls your topside opponent in and closer to you.

3: Pushing him to the side, puts him in that direction.

4: Pushing him back again, to continue the circle.

5: Pushing him to the side, puts him in that direction.

Step 1 may push a topside, punching opponent back and away from you. Step 2 may pull him down and closer to you where you can also interfere with his power punching, but also grab his head, neck for more moves. You might hook punch his head and neck. Steps 3 and 5 take him off to the sides, off his plan for the moment. Of course the opponent will fight you and these moves, so you must build as much power as possible to manipulate him.

The exercise with this is to manipulate a trainer, or maneuver some sort of weighted bag.

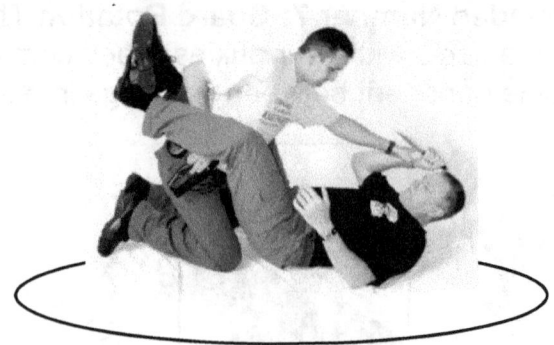

Hook ankles. Push feet out, which tightens the leg grip/vice on the opponent.

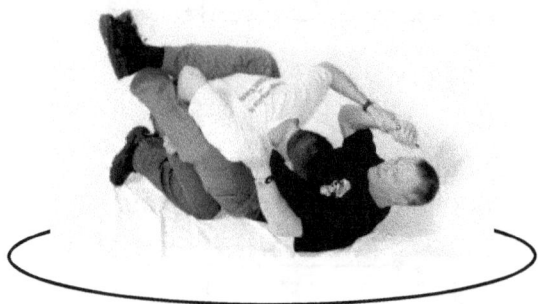

Grounded Number 8: The Common Roll Sometimes on the ground, you need to...rollover. Face up to face down. Face down to face up. Face up to face down to face up again, a foot or two from where you were. The worry for me is "giving up your back," for a potential choke, but that is when one engages in a very close quarters ground fight. The roll may be used when the opponent is, say, standing before you trying to stab you with a bayonetted rifle. Or, kicking down at you. Or you are evading gunfire. Not everything on the ground is an MMA fight.

Curve your shoulders, push with your foot and roll. Later there are exercises involving rolling over and getting up as fast as possible.

For an class exercise, a trainer can call out a rollover direction. A trainee rolls.

Grounded Number 9: The Scissor Kick Rollover Several situations on the ground call for this type of rollover. The face-up to face-down or face-down to face-up switch can be quickly and somewhat powerfully achieved by dynamically scissoring your legs and flipping your torso. Also move your arm to allow the turn.

Often your arm is holding an opponent's body part. A scissor kick can help you get out from under someone.

A trainer can call out a kick rollover direction.

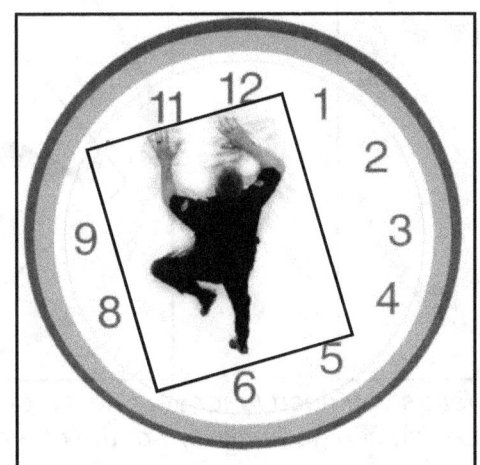

Grounded Number 10: Side Rotation, The Curly Shuffle 1/8, 1/4, 1/2, 3/4 or full circling. At times while training on the ground, I discovered I needed to lift my hip off the ground somewhat, put my weight on my shoulder and, or upper arm and with the use of my lower legs...circle. I called it the *Curly Shuffle* after Curly of the world famous "three Stooges." Frequently, Curly would drop to the floor and slip into a circle just like this.

This covers more "outer rim" circular space that confined the hip, small of back and belly pivots. A trainer can call out a clockwise or counter-clockwise direction.

Lift torso a bit, swing legs, pivot up on the shoulder and/or upper arm.

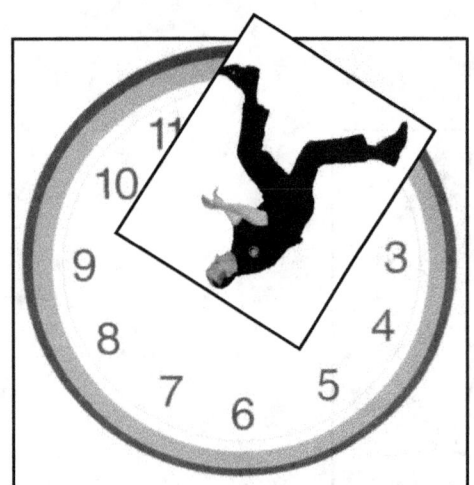 

Circle as far as you need to for your next movement.
Of course you can go backward or the other way.

The original Curly, caught in the shuffle!

Grounded Number 11: Back Pivot Rotation When suddenly on your back you may have to pivot to keep your feet aimed at or kicking at an opponent. The opponent could be standing, knee high or grounded. You push with your feet, a foot, elbow or elbows, hand or hands.

There is a lot of surface to your back, a lot of area to rub/friction against the ground as opposed to using your hip as a pivot point (shown next). From the back position you can bicycle pump your feet for thrusting kicks. A trainer can circle a trainee as the trainee spins on his back. The trainer can then hold a kicking shield for the trainee to kick.

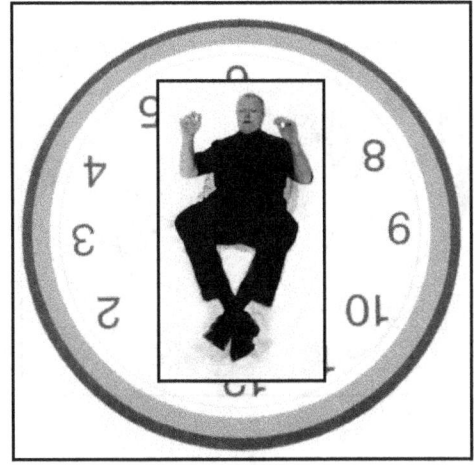

1: You are down on your back.

2: You pivot versus an enemy.

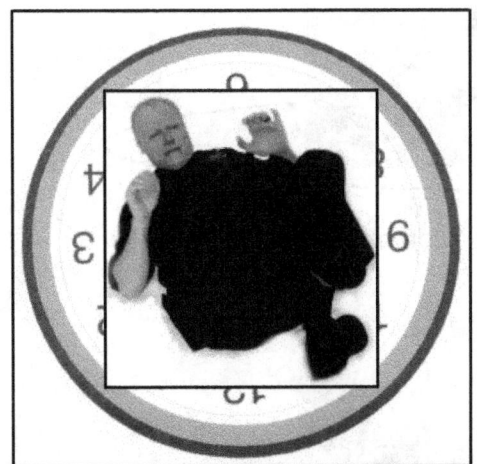

3: You have "bicycle pump," thrust kick options to temporarily fend off an opponent.

Grounded Number 12: Hip Pivot Rotation When suddenly on your back you may have to pivot to keep your feet aimed or kicking at an opponent. The opponent could be standing, knee high or grounded. To avoid the friction of the prior back pivot, you can roll to a hip. This makes for a fast turn. You can bicycle pump your feet for thrusting kicks. A trainer can circle a trainee as the trainee spins on his hip. The trainer can then hold a kicking shield for the trainee to kick.

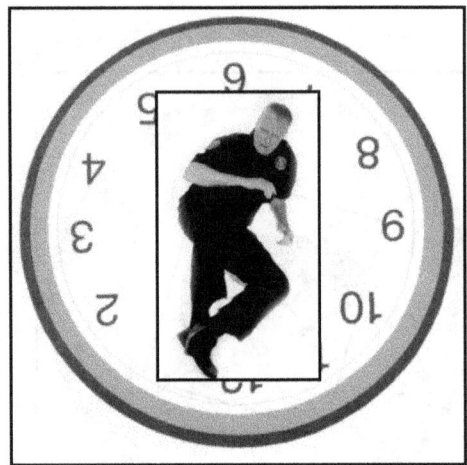 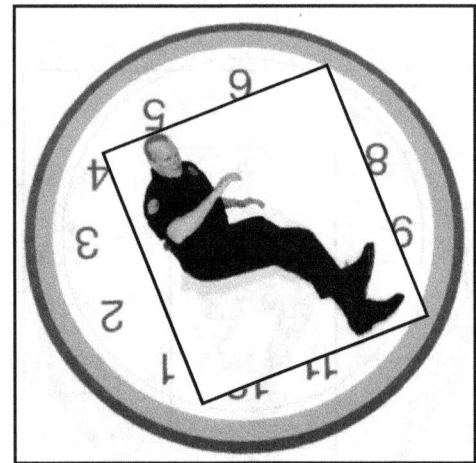

Jim McCann hip rotates atop an opponent.

Basic Unarmed and Armed Ground Maneuvers

Grounded Number 1: The Ready Position
- do unarmed or with weapons sheathed, holstered or mounted.
- do while drawing a knife, a stick, a pistol, a rifle.
- do with weapons presented.

Grounded Number 2: The Shrimp, or Hip Escape
- do unarmed or with weapons sheathed, holstered or mounted.
- do while drawing a knife, a stick, a pistol, a rifle.
- do with weapons presented.

Grounded Number 3: The Shoulder Walk
- do unarmed or with weapons sheathed, holstered or mounted.
- do while drawing a knife, a stick, a pistol, a rifle.
- do with weapons presented.

Grounded Number 4: The Bucking Bridge
- do unarmed or with weapons sheathed, holstered or mounted.
- do while drawing a knife, a stick, a pistol, a rifle.
- do with weapons presented.

Grounded Number 5: The Fishtails (Head and Torso)
- do unarmed or with weapons sheathed, holstered or mounted.
- do while drawing a knife, a stick, a pistol, a rifle.
- do with weapons presented.

Grounded Number 6: The Sit-Up
- do unarmed or with weapons sheathed, holstered or mounted
- do while drawing a knife, a stick, a pistol, a rifle
- do with weapons presented

Grounded Number 7: Guard Rotation
- do unarmed or with weapons sheathed, holstered or mounted.
- do while drawing a knife, a stick, a pistol, a rifle.
- do with weapons presented.

Grounded Number 8: The Common Rollover
- do unarmed or with weapons sheathed, holstered or mounted.
- do while drawing a knife, a stick, a pistol, a rifle.
- do with weapons presented.

Grounded Number 9: The Scissor Kick Rollover
- do unarmed or with weapons sheathed, holstered or mounted.
- do while drawing a knife, a stick, a pistol, a rifle.
- do with weapons presented.

Grounded Number 10: Side Rotation, The Curly Shuffle
- do unarmed or with weapons sheathed, holstered or mounted.
- do while drawing a knife, a stick, a pistol, a rifle.
- do with weapons presented.

Grounded Number 11: Back Pivot Rotation
- do unarmed or with weapons sheathed, holstered or mounted.
- do while drawing a knife, a stick, a pistol, a rifle.
- do with weapons presented.

Grounded Number 12: Hip Pivot Rotation
- do unarmed or with weapons sheathed, holstered or mounted.
- do while drawing a knife, a stick, a pistol, a rifle.
- do with weapons presented.

Segment 7: Universal Get-Up Maneuvers

Getting up. The basic tips for simply getting up off the floor/ground are found in all safety and medical advice.

    1: Get onto all fours.

    2: Bring the strong leg forward, knee bent, opposite hand on the floor for balance.

    3: Lift up, placing both hands on the front quad.

    4: Turn the back toes under and push your hands into the quad, using the strength of the thigh and upper body to push back to a standing position.

And also if empty-handed, there's this quick, classic pop up. Even if you are popping up from a topside ground fight. Your hand or hands my post/push right on the opponent.

However hand, stick, knife and gun combat puts us down on the ground in awkward positions. Typically the professional gun instructor suggests that you use the same steps that got you down there, in reverse to get back up.

To get back up, experts say you can often reverse the steps you took to go down.

The US Army also teaches another classic "hip-heist, sit-out," get-up. A "technical stand-up," as many call it. Raise up to clear the hip. Move the leg and foot back. Stabilize foot. Stand.

Posting is using a limb to get up. The "post" could be a fist, palm, forearm or elbow. Perhaps a knee. Be careful where you post if using the opponent's body. He or she may be naturally moving under you, throwing your posting off as well as your balance. They may also move or sweep your limb off their body with their arms or legs.

Posting with firearms, with the ups and downs of gun fight positioning, experts suggest to keep the barrel aimed at the enemy, along with your eyes.

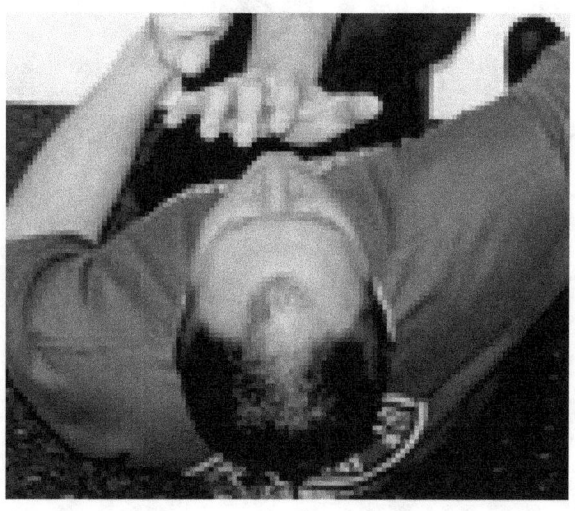 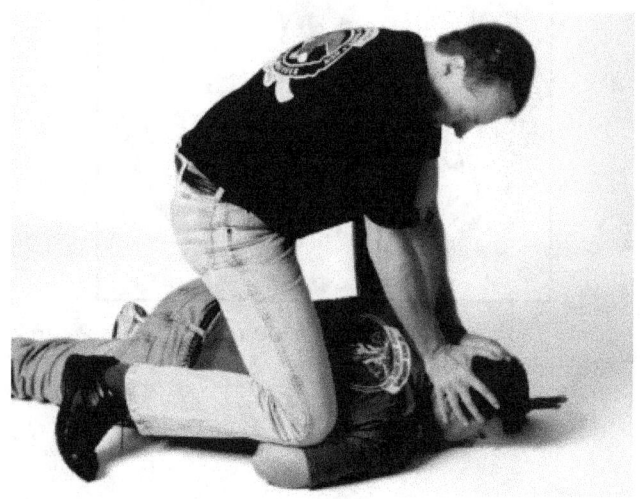

Getting Up and Run Off Exercise Number 1: The trainee lays flat face down. The trainer calls out a number and the trainee must pop up and run off the clock on that number.

Trainee is face down.

Trainee jumps up, runs off on the assigned number.

Getting Up and Run Off Exercise Number 2: The trainee lays flat, face up. The trainer calls out a number and the trainee must pop up and run off the clock at that number.

Trainee is face up.

Trainee jumps up, runs off on the assigned number.

Pressed against the wall. Here, Scott Pedersen demonstrates the wall-climb. It is essentially like the ground, shoulder walk. Easy to learn.

Get-Up, Unarmed and Armed Exercises

Face Down, Getting Up and Run Off Exercise Number 1:
- do unarmed or with weapons sheathed, holstered or mounted.
- do while drawing a knife, a stick, a pistol, a rifle.
- do with weapons presented.

Face Up, Getting Up and Run Off Exercise Number 2:
- do unarmed or with weapons sheathed, holstered or mounted.
- do while drawing a knife, a stick, a pistol, a rifle.
- do with weapons presented.

Wall Climb:
- do unarmed or with weapons sheathed, holstered or mounted.
- do while drawing a knife, a stick, a pistol, a rifle.
- do with weapons presented.

Segment 8: Obstacle Courses

"You're gonna run!"
"You're gonna crawl!"
"You're gonna climb!"
"You're gonna jump!"
"You're gonna zig!"
"You're gonna zag"

You can step around in one or two steps. You can crawl around an inch or two, or three. You move a bit on your knees. You can even walk and sprint. But can you chase, can you flee and escape, can you cover geography and all that it entails? With or without weapons. This is where the obstacle course comes in.

Historians explain that obstacle courses have been around since inception of militaries and the realization that troops need to move fast and furiously, quiet or loud, over specific terrain of battlefields and smaller missions.

I think you would be hard pressed to find anyone that isn't familiar with the concept of an obstacle course these days. They have been featured in movies and TV shows. The "ninja" course shows are on television all over the world. There was also a rise in the Tough Mudder/Spartan races throughout the world. Police and fire also use them for pass/fail hiring tests. Common sense tells us why.

(The following is military manual advice on obstacle courses, right out of many government manuals.)

Obstacle Courses
"Physical performance and success in combat may depend on a soldier's ability to perform skills like those required on the obstacle course. For this reason, and because they help develop and test basic motor skills, obstacle courses are valuable for physical training.

There are two types of obstacle courses--conditioning and confidence. The conditioning course has low obstacles that must be negotiated quickly. Running the course can be a test of the soldier's basic motor skills and physical condition. After soldiers receive instruction and practice the skills, they run the course against time.

A confidence course has higher, more difficult obstacles than a conditioning course. It gives soldiers confidence in their mental and physical abilities and cultivates their spirit of daring. Soldiers are encouraged, but not forced, to go through it. Unlike conditioning courses, confidence courses are not run against time.

NONSTANDARD COURSES AND OBSTACLES

Commanders may build obstacles and courses that are nonstandard (that is, not covered in this manual) in order to create training situations based on their unit's mission.

When planning and building such facilities, designers should, at a minimum, consider the following guidance:

- Secure approval from the local installation's commander
- Prepare a safety and health-risk assessment to support construction of each obstacle
- Coordinate approval for each obstacle with the local or supporting safety office
- Keep a copy of the approval in the permanent records
- Monitor and analyze all injuries
- Inspect all existing safety precautions on-site to verify their effectiveness
- Review each obstacle to determine the need for renewing its approval

SAFETY PRECAUTIONS

Instructors must always be alert to safety. They must take every precaution to minimize injuries as soldiers go through obstacle courses. Soldiers must do warm-up exercises before they begin. This prepares them for the physically demanding tasks ahead and helps minimize the chance of injury. A cool-down after the obstacle course is also necessary, as it helps the body recover from strenuous exercise.

Commanders should use ingenuity in building courses, making good use of streams, hills, trees, rocks, and other natural obstacles. They must inspect courses for badly built obstacles, protruding nails, rotten logs, unsafe landing pits, and other safety hazards.

There are steps which designers can take to reduce injuries. For example, at the approach to each obstacle, they should post an instruction board or sign with text and pictures showing how to negotiate it. Landing pits for jumps or vaults, and areas under or around obstacles where soldiers may fall from a height, should be filled with loose sand or sawdust. All landing areas should be raked and refilled before each use. Puddles of water under obstacles can cause a false sense of security. These could result in improper landing techniques and serious injuries. Leaders should postpone training on obstacle courses when wet weather makes them slippery.

Units should prepare their soldiers to negotiate obstacle courses by doing conditioning exercises beforehand. Soldiers should attain an adequate level of conditioning before they run the confidence course. Soldiers who have not practiced the basic skills or run the conditioning course should not be allowed to use the confidence course.

Instructors must explain and demonstrate the correct ways to negotiate all obstacles before allowing soldiers to run them. Assistant instructors should supervise the negotiation of higher, more dangerous obstacles. The emphasis is on avoiding injury. Soldiers should practice each obstacle until they are able to negotiate it. Before they run the course against time, they should make several slow runs while the instructor watches and makes needed corrections. Soldiers should never be allowed to run the course against time until they have practiced on all the obstacles.

CONDITIONING OBSTACLE COURSES

If possible, an obstacle course should be shaped like a horseshoe or figure eight so that the finish is close to the start. Also, signs should be placed to show the route.

A course usually ranges from 300 to 450 yards and has 15 to 25 obstacles that are 20 to 30 yards apart. The obstacles are arranged so that those which exercise the same groups of muscles are separated from one another.

The obstacles must be solidly built. Peeled logs that are six to eight inches wide are ideal for most of them. Sharp points and corners should be eliminated, and landing pits for jumps or vaults must be filled with sand or sawdust. Courses should be built and marked so that soldiers cannot sidestep obstacles or detour around them. Sometimes, however, courses can provide alternate obstacles that vary in difficulty.

Each course should be wide enough for six to eight soldiers to use at the same time, thus encouraging competition. The lanes for the first few obstacles should be wider and the obstacles easier than those that follow. In this way, congestion is avoided and soldiers can spread out on the course. To minimize the possibility of falls and injuries due to fatigue, the last two or three obstacles should not be too difficult or involve high climbing.

Trainers must always be aware that falls from the high obstacles could cause serious injury. Soldiers must be in proper physical condition, closely supervised, and adequately instructed.

The best way for the timer to time the runners is to stand at the finish and call out the minutes and seconds as each soldier finishes. If several watches are available, each wave of soldiers is timed separately. If only one watch is available, the waves are started at regular intervals such as every 30 seconds. If a soldier fails to negotiate an obstacle, a previously determined penalty is imposed.

When the course is run against time, stopwatches, pens, and a unit roster are needed. Soldiers may run the course with or without individual equipment.

CONFIDENCE OBSTACLE COURSES

Confidence courses can develop confidence and strength by using obstacles that train and test balance and muscular strength. Soldiers do not negotiate these obstacles at high speed or against time. The obstacles vary from fairly easy to difficult, and some are high. For these, safety nets are provided. Soldiers progress through the course without individual equipment. Only one soldier at a time negotiates an obstacle unless it is designed for use by more than one.

Confidence courses should accommodate four platoons, one at each group of six obstacles. Each platoon begins at a different starting point. In the example below, colors are used to group the obstacles. Any similar method may be used to spread a group over the course. Soldiers are separated into groups of 8 to 12 at each obstacle. At the starting signal, they proceed through the course.

Soldiers may skip any obstacle they are unwilling to try. Instructors should encourage fearful soldiers to try the easier obstacles first. Gradually, as their confidence improves, they can take their places in the normal rotation. Soldiers proceed from one obstacle to the next until time is called. They then assemble and move to the next group of obstacles.

Rules for the Course
Supervisors should encourage, but not force, soldiers to try every obstacle. Soldiers who have not run the course before should receive a brief orientation at each obstacle, including an explanation and demonstration of the best way to negotiate it. Instructors should help those who have problems. Trainers and soldiers should not try to make obstacles more difficult by shaking ropes, rolling logs, and so forth. Close supervision and common sense must be constantly used to enhance safety and prevent injuries."

The following are examples of various obstacle courses.

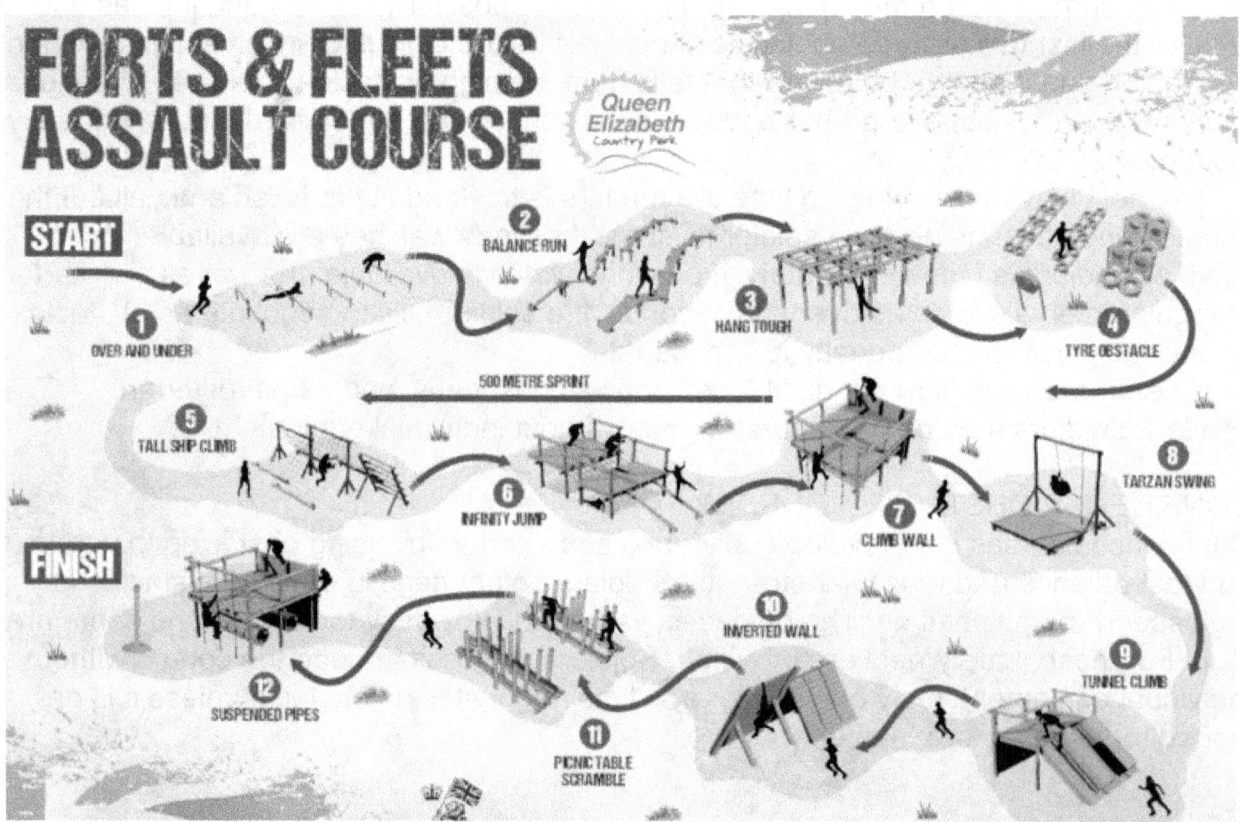

$13,000 from Hammacher Schlemmer.

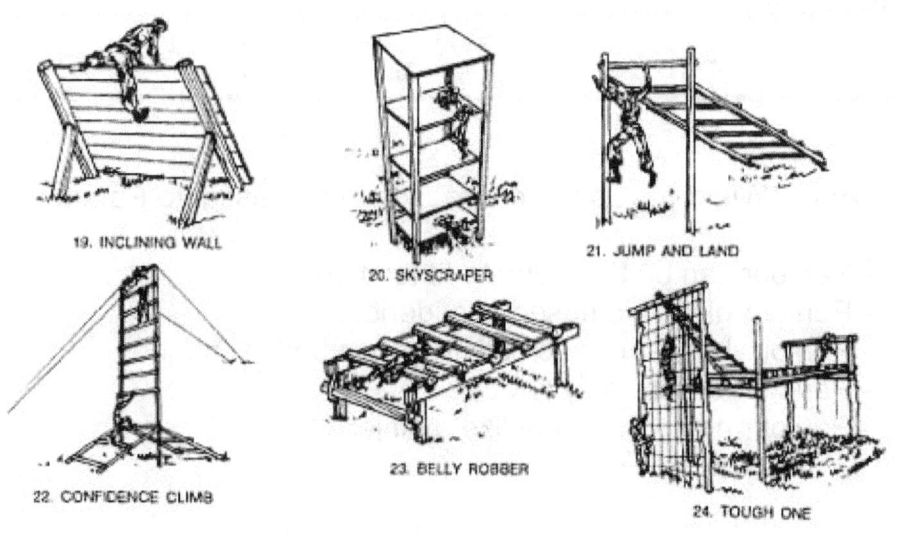

Does this lead us to Parkour?
World Freerunning and Parkour Federation says, "The word "Parkour" comes from the French "parcours," which literally means, "the way through," or "the path. "What we now all know as "Parkour" with a "k" had its origins in a training program for French Special

Forces known as "Parcours du combattant", or "The Path of the Warrior." It was David Belle, a French dude, son of a Parcours Warrior and the inventor of Parkour, who changed the "c" to a "k" and, along with his comrades, the Yamakazi, began the world-wide movement you are now officially a part of and which also includes the phenomenon known as Freerunning (confused yet? Don't give up! You're almost there!)

According to the strictest definition, Parkour is the act of moving from point "a" to point "b" using the obstacles in your path to increase your efficiency. Sounds like a fun game, right? A basic repertoire of moves developed over the years, like the "tic-tac", the "kong vault" and the "gap jump" that make Parkour immediately recognizable to most people who see it, even if they don't know what it's called!"

Most people were exposed to the idea when watching the movie "Casino Royale," but we all have seen exciting acrobatics in films for decades. While I don' think citizens, police and the military should be leaping buildings in single bounds, learning some common sense maneuvers around common sense things is smart.

Practitioner Adam Wik gives us 5 more reasons to investigate Parkour:

- Parkour can be the ultimate fitness plan.
- Parkour gives increased confidence.
- Parkour brings more creativity and a better attitude.
- Parkour is extremely fun.
- Parkour makes you feel like a ninja.

Obstacle Course Work

Experience obstacle courses that make you:
- run.
- crawl.
- climb.
- jump.
- zig.
- zag.
- without weapons.
- with holstered, sheathed and shouldered weapons.
- with weapons drawn.

Ankle Breaks

Through the 1970s to the 1990s, I noticed a tripping accident was fairly common in line operations, police work. Ankle juries. Line Ops is often synonymous with chasing people and dashing to active crime scenes. Running. Running over the urban, suburban and rural terrain, and looking far off, and not down on the objects and contours on the ground before you.

So that's a bad enough invitation to an ankle twist or break or fall. But another thing I noticed and not just in my agency and the surrounding agencies, and then nationally – another unique accident. Police cars parking hurriedly beside curbs and other crap, and officers bailing out of cars, looking off to problem people and places, and catching their ankles on stuff. Sprain, Or break. Or fall. (I also heard similar stories from the military, where by the nature of what they do in total, the ankles are weak links in action.)

In one week, we had CID captain bail out and break an ankle, and a veteran patrol officer bail out and break his ankle, because of curbs. Both were passengers by the way. So the while driver could guess-see where he was going, the passenger was stuck with what he got over on his side. The captain's gun was out. The officer was pulling his gun. Think of the residual mess a discharge would have made. Could have made. There are a number of discharges each year with falls. Fingers off the trigger!

That strange week was when I began to take notice of the problem. Many moons ago. This type of thing, a car bail-out, least of all a foot chase could happen to any ambitious person, gun or not, police or not. Military or not.

Look around! But one thing to do in preparation is to develop more resilient ankles. Not just calf raises up and down, but rotating your foot and rocking it side to side under a weight pressure. Leg work outs, even running create a better ankle to withstand surprises in the future. (I might add here that the two cases I mentioned above…neither worked out.)

Since all that, I take a quick look down. Or look fast and remember where I will be when I pull up somewhere. Sometimes it could just be junk, muck or a giant puddle out there. Warn your partner if you have one.

Segment 9: Your Personal, Class and, or Seminar Workout List
Experiment with the "do" list. See if you can or should do the "do" list.

Footwork and Maneuvering Exercise Number 1
Stand Straight and Step to "Bladed" Ready Footwork:
- step the right foot back and to some degree, crouch.
    * do unarmed or with weapons sheathed, holstered or mounted.
    * do while drawing and holding a knife.
    * do while drawing and holding a stick, or a flashlight.
    * do while drawing and holding a pistol.
    * do while unshouldering/lifting and holding a rifle.

- step the left foot back and to some degree, crouch.
    * do unarmed or with weapons sheathed, holstered or mounted.
    * do while drawing and holding a knife.
    * do while drawing and holding a stick or a flashlight.
    * do while drawing and holding a pistol.
    * do while unshouldering/lifting and holding a rifle.

Footwork and Maneuvering Exercise Number 2
One foot, In and Out Footwork:
- left foot on axis, right foot goes from 2 to 4 and back.
    * do unarmed or with weapons sheathed, holstered or mounted.
    * do while drawing and holding a knife.
    * do while drawing and holding a stick or a flashlight.
    * do while drawing and holding a pistol.
    * do while unshouldering/lifting and holding a rifle.

- right foot on axis, left foot goes from 10 to 8 and back.
    * do unarmed or with weapons sheathed, holstered or mounted.
    * do while drawing and holding a knife.
    * do while drawing and holding a stick or a flashlight.
    * do while drawing and holding a pistol.
    * do while unshouldering/lifting and holding a rifle.

Footwork and Maneuvering Exercise Number 3
V Footwork: Bladed with right foot forward.
- run the right foot back to the axis and shoot the left foot to 10 o'clock.
    * do unarmed or with weapons sheathed, holstered or mounted.
    * do while drawing and holding a knife.
    * do while drawing and holding a stick or a flashlight.
    * do while drawing and holding a pistol.
    * do while unshouldering/lifting and holding a rifle.

V Footwork: Bladed with left foot forward.
- bring the left foot back to the axis and shoot the right foot to 1 o'clock.
    * do unarmed or with weapons sheathed, holstered or mounted.
    * do while drawing and holding a knife.
    * do while drawing and holding a stick or a flashlight.
    * do while drawing and holding a pistol.
    * do while unshouldering/lifting and holding a rifle.

V then turn/pivot inward at the 10 and 2 clock points.
    * do unarmed or with weapons sheathed, holstered or mounted.
    * do while drawing and holding a knife.
    * do while drawing and holding a stick or a flashlight.
    * do while drawing and holding a pistol.
    * do while unshouldering/lifting and holding a rifle.

Footwork and Maneuvering Exercise Number 4
Upside Down V Footwork: Start bladed with right foot back. Other foot on the axis. The goal is the 8 and 4 o'clock points
- bring the right foot forward to the axis and shoot the left foot back to 7 o'clock, creating the reverse, upside down V.
    * do unarmed or with weapons sheathed, holstered or mounted.
    * do while drawing and holding a knife.
    * do while drawing and holding a stick or a flashlight.
    * do while drawing and holding a pistol.
    * do while unshouldering/lifting and holding a rifle.
    * do the pattern with the left foot back.

Footwork and Maneuvering Exercise Number 5
Shuffle Footwork. Start bladed. One foot moves near the other, then that other foots shoots out. When you do the return motion, you are covering both sides of the clock.
- 1 to axis to 7 and back.
    * do unarmed or with weapons sheathed, holstered or mounted.
    * do while drawing and holding a knife.
    * do while drawing and holding a stick or a flashlight.
    * do while drawing and holding a pistol.
    * do while unshouldering/lifting and holding a rifle.

- 2 to axis to 8 and back.
    * do unarmed or with weapons sheathed, holstered or mounted.
    * do while drawing and holding a knife.
    * do while drawing and holding a stick or a flashlight.
    * do while drawing and holding a pistol.
    * do while unshouldering/lifting and holding a rifle.

- 3 to axis to 9 and back.
    * do unarmed or with weapons sheathed, holstered or mounted.
    * do while drawing and holding a knife.
    * do while drawing and holding a stick or a flashlight.
    * do while drawing and holding a pistol.
    * do while unshouldering/lifting and holding a rifle.

- 4 to axis to 10 and back.
    * do unarmed or with weapons sheathed, holstered or mounted.
    * do while drawing and holding a knife.
    * do while drawing and holding a stick or a flashlight.
    * do while drawing and holding a pistol.
    * do while unshouldering/lifting and holding a rifle.

- 5 to axis to 11 and back.
    * do unarmed or with weapons sheathed, holstered or mounted.
    * do while drawing and holding a knife.
    * do while drawing and holding a stick or a flashlight.
    * do while drawing and holding a pistol.
    * do while unshouldering/lifting and holding a rifle.

- 6 to axis to 12 and back.
    * do unarmed or with weapons sheathed, holstered or mounted.
    * do while drawing and holding a knife.
    * do while drawing and holding a stick or a flashlight.
    * do while drawing and holding a pistol.
    * do while unshouldering/lifting and holding a rifle.

- 12 to axis to 6 and back.
    * do unarmed or with weapons sheathed, holstered or mounted.
    * do while drawing and holding a knife.
    * do while drawing and holding a stick or a flashlight.
    * do while drawing and holding a pistol.
    * do while unshouldering/lifting and holding a rifle.

Footwork and Maneuvering Exercise Number 6
Big Circle and Shrink Down to Small Pivot Circle Footwork.
    Clockwise
        * do unarmed or with weapons sheathed, holstered or mounted.
        * do while drawing and holding a knife.
        * do while drawing and holding a stick or a flashlight.
        * do while drawing and holding a pistol.
        * do while unshouldering/lifting and holding a rifle.

Counter-clockwise
- * do unarmed or with weapons sheathed, holstered or mounted.
- * do while drawing and holding a knife.
- * do while drawing and holding a stick or a flashlight.
- * do while drawing and holding a pistol.
- * do while unshouldering/lifting and holding a rifle.

Increasing smaller circle, clockwise or counter-clockwise.
- * do unarmed or with weapons sheathed, holstered or mounted.
- * do while drawing and holding a knife.
- * do while drawing and holding a stick or a flashlight.
- * do while drawing and holding a pistol.
- * do while unshouldering/lifting and holding a rifle.

Pivot footwork, pivot on right foot in clockwise circle.
- * do unarmed or with weapons sheathed, holstered or mounted.
- * do while drawing and holding a knife.
- * do while drawing and holding a stick or a flashlight.
- * do while drawing and holding a pistol.
- * do while unshouldering/lifting and holding a rifle.

Pivot footwork, pivot on left foot in a counter-clockwise circle.
- * do unarmed or with weapons sheathed, holstered or mounted.
- * do while drawing and holding a knife.
- * do while drawing and holding a stick or a flashlight.
- * do while drawing and holding a pistol.
- * do while unshouldering/lifting and holding a rifle.

Footwork and Maneuvering Exercise Number 7
Walking Lanes with Sharp Turns.
- * do unarmed or with weapons sheathed, holstered or mounted.
- * do while drawing and holding a knife.
- * do while drawing and holding a stick or a flashlight.
- * do while drawing and holding a pistol.
- * do while unshouldering/lifting and holding a rifle.

Footwork and Maneuvering Exercise Number 8
Run Off the Clock.
- * do unarmed or with weapons sheathed, holstered or mounted.
- * do while drawing and holding a knife.
- * do while drawing and holding a stick or a flashlight.
- * do while drawing and holding a pistol.
- * do while unshouldering/lifting and holding a rifle.

Footwork and Maneuvering Exercise Number 9
The Knee Stretch Exercise.
* do unarmed or with weapons sheathed, holstered or mounted.
* do with weapons presented.

Footwork and Maneuvering Exercise Number 10
Knee Squat Down.
* do unarmed or with weapons sheathed, holstered or mounted.
* do while drawing and holding a knife.
* do while drawing and holding a stick or a flashlight.
* do while drawing and holding a pistol.
* do while unshouldering/lifting and holding a rifle.

Footwork and Maneuvering Exercise Number 11
Knee-Step Off.
* do unarmed or with weapons sheathed, holstered or mounted.
* do while drawing and holding a knife.
* do while drawing and holding a stick or a flashlight.
* do while drawing and holding a pistol.
* do while unshouldering/lifting and holding a rifle.

Footwork and Maneuvering Exercise Number 12
Knee-Step Off to a Knee Up.
* do unarmed or with weapons sheathed, holstered or mounted.
* do while drawing and holding a knife.
* do while drawing and holding a stick or a flashlight.
* do while drawing and holding a pistol.
* do while unshouldering/lifting and holding a rifle.

Footwork and Maneuvering Exercise Number 12
The Side-to-Side Torso Glide.
* do unarmed or with weapons sheathed, holstered or mounted.
* do while drawing and holding a knife.
* do while drawing and holding a stick or a flashlight.
* do while drawing and holding a pistol.
* do while unshouldering/lifting and holding a rifle.

Footwork and Maneuvering Exercise Number 13
Get Down 1: Standing (to a Knee or to two knees) to Prone with Two Hands.
* do unarmed or with weapons sheathed, holstered or mounted.

Footwork and Maneuvering Exercise Number 14
Get Down 2: Standing to a Knee (or to Two Knees) to Prone with One Hand.
* do unarmed or with weapons sheathed, holstered or mounted.
* do while drawing and holding a knife.
* do while drawing and holding a stick or a flashlight.
* do while drawing and holding a pistol.
* do while unshouldering/lifting and holding a rifle.

Footwork and Maneuvering Exercise Number 15
Get Down 3: Standing to a Rolling Side Drop.
* do unarmed or with weapons sheathed, holstered or mounted.
* do while drawing and holding a knife.
* do while drawing and holding a stick or a flashlight.
* do while drawing and holding a pistol.
* do while unshouldering/lifting and holding a rifle.

Footwork and Maneuvering Exercise Number 16
Get Down 4: Dropping Supine.
* do unarmed or with weapons sheathed, holstered or mounted.
* do while drawing and holding a knife.
* do while drawing and holding a stick or a flashlight.
* do while drawing and holding a pistol.
* do while unshouldering/lifting and holding a rifle.

Footwork and Maneuvering Exercise Number 17:
Ground Maneuver 1: The Shrimp, or Hip Escape.
* do unarmed or with weapons sheathed, holstered or mounted.
* do while drawing and holding a knife.
* do while drawing and holding a stick or a flashlight.
* do while drawing and holding a pistol.
* do while unshouldering/lifting and holding a rifle.

Footwork and Maneuvering Exercise Number 18:
Ground Maneuver 2: The Shoulder Walk.
* do unarmed or with weapons sheathed, holstered or mounted.
* do while drawing and holding a knife.
* do while drawing and holding a stick or a flashlight.
* do while drawing and holding a pistol.
* do while unshouldering/lifting and holding a rifle.

Footwork and Maneuvering Exercise Number 19:
Ground Maneuver 3: The Bucking Bridge.
* do unarmed or with weapons sheathed, holstered or mounted.
* do while drawing and holding a knife.
* do while drawing and holding a stick or a flashlight.
* do while drawing and holding a pistol.
* do while unshouldering/lifting and holding a rifle.

Footwork and Maneuvering Exercise Number 20:
Ground Maneuver 4: The Fishtails (Head and Torso).
* do unarmed or with weapons sheathed, holstered or mounted.
* do while drawing and holding a knife.
* do while drawing and holding a stick or a flashlight.
* do while drawing and holding a pistol.
* do while unshouldering/lifting and holding a rifle.

Footwork and Maneuvering Exercise Number 21:
Ground Maneuver 5: The Sit-Up.
* do unarmed or with weapons sheathed, holstered or mounted.
* do while drawing and holding a knife.
* do while drawing and holding a stick or a flashlight.
* do while drawing and holding a pistol.
* do while unshouldering/lifting and holding a rifle.

Footwork and Maneuvering Exercise Number 21:
Ground Maneuver 6: Guard Rotation.
* do unarmed or with weapons sheathed, holstered or mounted.
* do while drawing and holding a knife.
* do while drawing and holding a stick or a flashlight.
* do while drawing and holding a pistol.
* do while unshouldering/lifting and holding a rifle.

Footwork and Maneuvering Exercise Number 22:
Ground Maneuver 7: The Common Rollover.
    * do unarmed or with weapons sheathed, holstered or mounted.
    * do while drawing and holding a knife.
    * do while drawing and holding a stick or a flashlight.
    * do while drawing and holding a pistol.
    * do while unshouldering/lifting and holding a rifle.

Footwork and Maneuvering Exercise Number 23:
Ground Maneuver 8: The Scissor Kick Rollover.
    * do unarmed or with weapons sheathed, holstered or mounted.
    * do while drawing and holding a knife.
    * do while drawing and holding a stick or a flashlight.
    * do while drawing and holding a pistol.
    * do while unshouldering/lifting and holding a rifle.

Footwork and Maneuvering Exercise Number 24
Face Down, Getting Up and Run Off Exercise Number 1:
    * do unarmed or with weapons sheathed, holstered or mounted.
    * do while drawing and holding a knife.
    * do while drawing and holding a stick or a flashlight.
    * do while drawing and holding a pistol.
    * do while unshouldering/lifting and holding a rifle.

Footwork and Maneuvering Exercise Number 25
Face Up, Getting Up and Run Off Exercise Number 2:
- do unarmed or with weapons sheathed, holstered or mounted.
- do while drawing a knife, a stick, a pistol, a rifle.
- do with weapons presented.
    * do unarmed or with weapons sheathed, holstered or mounted.
    * do while drawing and holding a knife.
    * do while drawing and holding a stick or a flashlight.
    * do while drawing and holding a pistol.
    * do while unshouldering/lifting and holding a rifle.

Footwork and Maneuvering Exercise Number 26
The Wall Climb:
- do unarmed or with weapons sheathed, holstered or mounted.
- do while drawing a knife, a stick, a pistol, a rifle.
- do with weapons presented.

Footwork and Maneuvering Exercise Number 27
Obstacle course work
* do unarmed or with weapons sheathed, holstered or mounted.
* do while drawing and holding a knife.
* do while drawing and holding a stick or a flashlight.
* do while drawing and holding a pistol.
* do while unshouldering/lifting and holding a rifle.

Footwork and Maneuvering Exercise Number 28
Ambush, Dodge, Evasion Drill
Remember back in the beginning of this book, we displayed the ten attacks? Head-head. Torso-torso. Leg-leg. Up-down. Straight low, straight high. The trainer used body motion and the trainer attacks with:
* Standing versus standing
    - a knife.
    - a stick.
    - fists and kick.

* Standing versus grounded
    - a knife.
    - a stick.
    - fists and kicks.

* From here, you can develop combat scenarios

Get these other martial books by Hock.

Find them at Amazon, Barnes & Nobles and at *www.ForceNecessary.com*